City Echoes: Lessons from New York

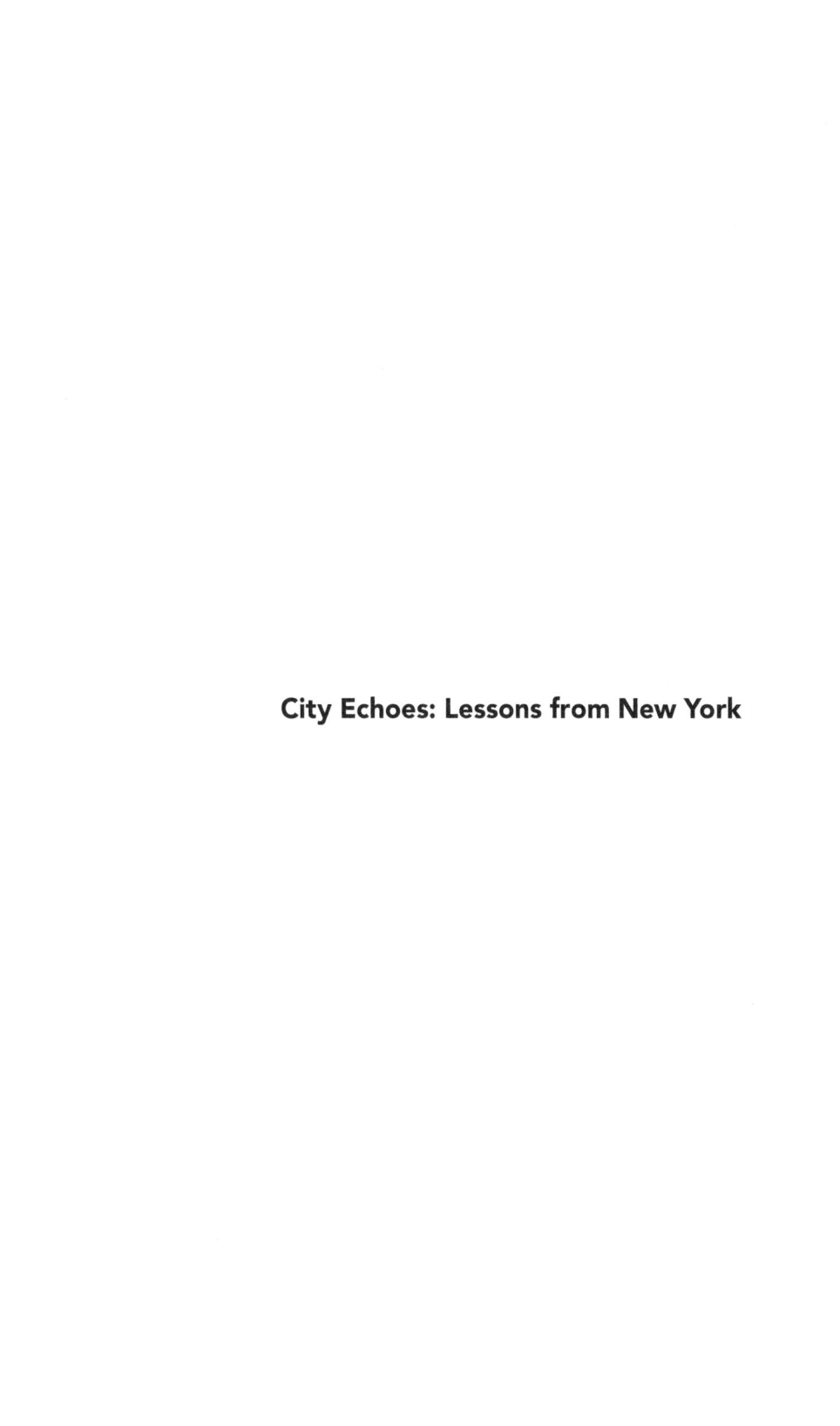

City Echoes: Lessons from New York

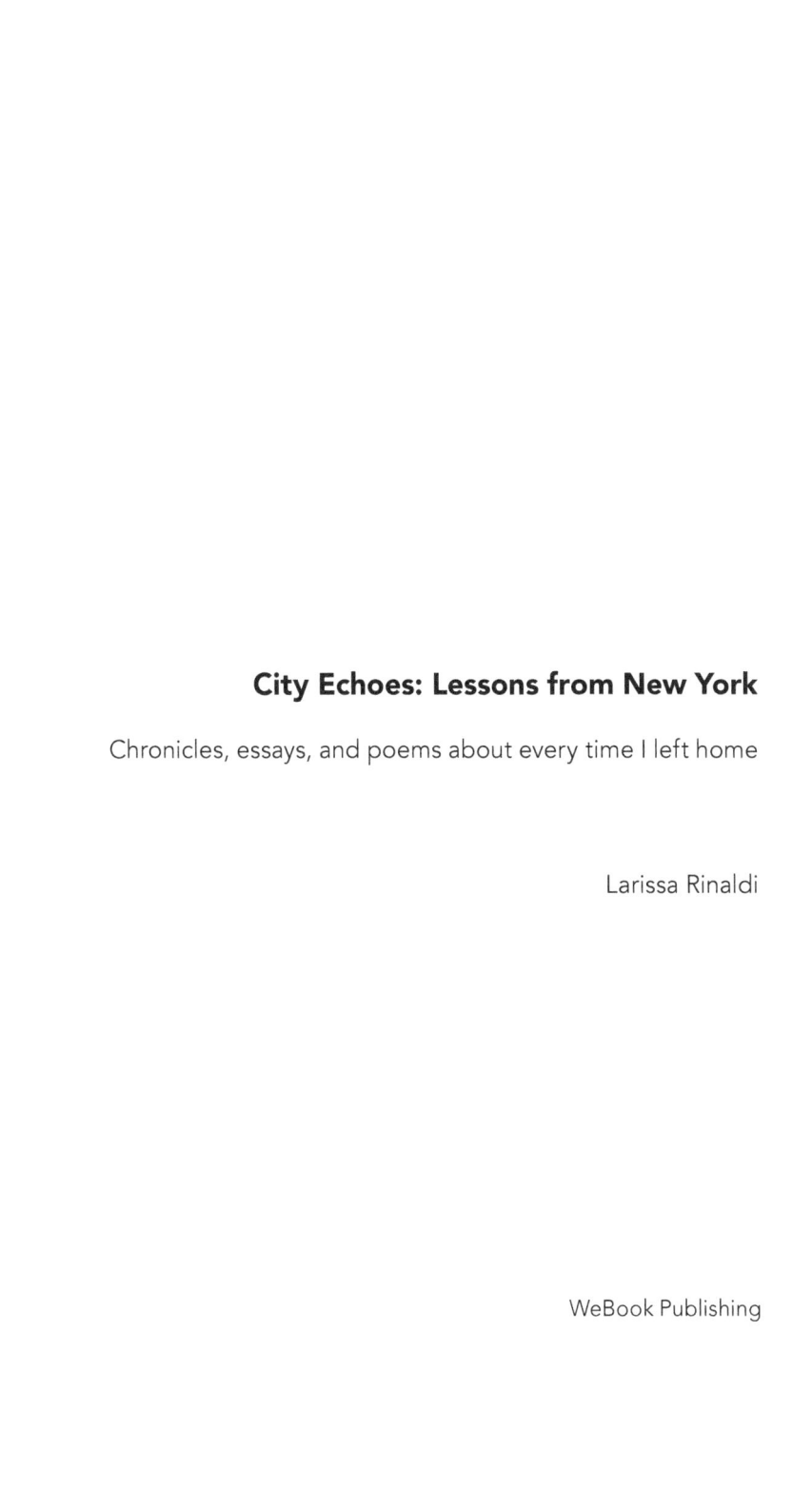

Chronicles, essays, and poems about every time I left home

Larissa Rinaldi

WeBook Publishing

Published by WeBook Publishing – Los Angeles, CA
All rights in the English language reserved.

No portion of this book may be copied, stored in recovery systems, or transferred by any means, whether electronic or mechanical, nor photocopied, recorded, or otherwise, without the author's and the Publisher's written permission.
This is a work of nonfiction based on the author's experiences adapting to a new country, language, culture, and life.

For information, please email info@webookpublishing.com

Copyright © 2024 Larissa Rinaldi
English Edition

ISBN: 979-979-89886684-6-6
LCCN: 2024908691
Written by Larissa Rinaldi
Translator: Laura Linn
Translation Revisions: Ana Silvani
Copy Editing: Ana Silvani
Cover Art: Marcela Abbade

Manufactured in the United States of America

Note:
Much care and technique were employed in editing this book. However, there can be no assurance that it will be free of minor typing errors, printing issues, or even conceptual ambivalence. In any such case, we ask that the issue be notified to our customer service at the e-mail address info@webookpublishing.com

---

"Writing a book is a long journey filled with challenges and discoveries over the past few years. I had the help of dear friends who watched me with care. I especially thank my friends who lent me some of their time and knowledge to help me take this manuscript out of my virtual drawer and into the real world. Thank you, Beatriz Abud, Gabriela Cuzzuol Ribeiro, Heloisa Villela, Luma Cabral, Nélida Costa, Yan Della Torre and Marcela Abbade."

Larissa Rinaldi

Table of Contents

8	Disclaimer
9	A book about self-discovery
11	**Who am I?**
13	What within me is me?
33	The Beginning of the Turnaround
43	The Spirals Of Life
51	The Arrogance of Writing About Life
71	**Skid Marks of Me on the Pavement**
73	A *carioca* from Rio
81	*Carioca* in SP
91	City Echoes: lessons from New York
115	**The End and the Beginning**
117	Post-pandemic texts
127	Your Future Image
141	I arrived here

Disclaimer

The following are writings and city echoes from 2020 to 2021. The author and the publisher have decided to maintain the text in its original form for this English edition to keep the beauty of the writer's feelings during such a dark moment in human history. Yes, we are about to take you back to the past, a very recent past that somehow already feels like ages ago.

A book about self-discovery

I arrived in New York in 2018 at 29 years old, right in the middle of Saturn's Return, at that phase in your life where you truly become an adult.

I left my comfort zone and wanted to return to it as quickly as possible, but it wasn't that simple. Becoming an adult is getting used to the (un)comfortable.

I came to live in the most diverse city in the world, where adaptation is too simple a word to describe the whirlwind of emotions that life here entails.

Before the pandemic, I was already in quarantine, like a spoiled child who wanted to keep hiding behind their mother's skirts.

In recent years, I had a lot of time to look inward. While Saturn pulled me towards reality, I wrote down what I was learning, like someone writing down lessons in a notebook. My lessons turned into my first book. Chronicles, essays, and poems about Everything I learned in New York.

Only those who have lived it know how the process of (re)discovering oneself professionally abroad truly feels. But hardly anyone truly knows. I am finding out as I go along.

Who am I?

What within me is me?

I wanted to create a poetic introduction for each part of my first published book.

So proud!

A book written and signed by Larissa Rinaldi, also known as me.

I never felt like I had a place in the world and thought about documenting it here. But what is the point of writing a brief text to explain what all the chronicles in this book seek to unravel:

What within me is me, and what are other people?

I came to the conclusion that it's best not to clarify or even introduce chapters. This is a book that journeys into the depths of the dark soul, with secret passages and returns to the beginning, like the setbacks in board games, with no guarantees of an exit.

It must be challenging to be a writer

Not wanting, at times, is normal.
No one wants the same thing every second of every day of their life.
Or who knows... maybe someone does.
I'm not that person,
despite wanting, sometimes,
even thinking it will last forever.
But then, life happens
And the desire to experience other things arises.
The problem must be this senseless need that rebels against the comfort of following paved roads

This inconvenient desire to see the world through my own lenses grows without asking for permission

Paving the way in the raw state of adult life, not asking for advice

Me by myself

Innocence, foolishness, or trauma

I've rambled on too much and lost focus, as usual.

The fact is, being a female writer must be challenging.

Everything is ephemeral except me

In 2020, everything was ephemeral except me
Not that I'm special
but the Self is not ephemeral
Neither mine nor yours
Even though our smallest facets are echoed ephemerally
In digital networks

Of 15-second videos
and meaningless photos

I am not ephemeral
I must exist
or else no one knows me

Yet, inside, I am deep with darkness, echo, and light within a unique, complex, and dead-end body.

What within me is me?

For some time, I've questioned what in me is:
1. my desire;
2. others projections onto me;
3. my projection to please others.

The subject is complex, and journeying into the depths of the soul is an irreversible path. Don't worry, or worry if you want. This is a matter between you and yourself, and I won't intervene; I will, just share my side of things.

I've been in therapy for years, and I'm beginning to accept the fact that some people just know what they want.

They lay out a plan and follow the path with much effort and few deviations. It's tough to live with this reality when you - I, in this case - always feel lost.

I am different. I've always found it hard to face myself, stand up for my desires, and ignore external expectations. Cowardice eats away, diminishes, and triggers terrible anxiety crises.

Then, therapy comes in; then, medication.

Bless the brave souls who inspire and support me, even if unknowingly, even if they don't understand why I choose the winding roads back to Santos (a reference to a famous Brazilian song released in 1969: "As Curvas da Estrada de Santos").

As for the rest, I'd like to hate as much as I think they hate me or ignore as much as they ignore me, because silence, silence is poison to the artist's soul, and that I cannot forgive.

Untitled

Everyone wants what they don't have, right? It's the biggest maxim in the world, the biggest cliché of humanity..

Wanting something? Don't have it.

Want to stop wanting? Have it.

Want to enjoy the single life? Get a girlfriend.

Want to live abroad? Leave and miss Brazil.

I don't know what happens to human minds for us to be like this.

Perhaps it's a fear of losing. I had an astrological chart done, andthe astrologer said I dislike showing emotions. Reading this book, you wouldn't say but try talking to me on the street. I'm an ice cube, not because I truly am, but because I'm good at hiding emotions.

(– Do you remember my cage? Have you reached this part of the book yet?)

The astrologer told me that there was a major rupture in my family when I was two years old. I, who thought my life was over when my parents separated when I was 6, completely ignored the death of my uncle, the eldest of the family, four years earlier.

I, who don't really like to feel (notice here that I've already given a hint of who I truly am, not just what I appear to be on the street), said I didn't know how that could affect my life, as I was so young. She, with that annoyingly patient expression typical of those who know something you don't, invited me to look around, even though I was small, even at two years old. How did my family cope after that death?

I don't know if it's true, but it's a family legend that my grandfather died a few years later. Cause of death: sadness.

I can't guarantee any of this. I could ask my father, but I imagine that this is a subject that brings a lot of pain, and I don't want to make my father suffer. I could ask my aunt, but I stopped talking to her in 2018. That leaves my cousin, who is now in Australia and uses the time difference as an excuse to disconnect. Marcela, my wife, occasionally, as if nonchalant, asks about the cousin who used to visit our house. I say I don't know and don't care, but she asks because I do care, but since she (cousin) doesn't care, it's okay to talk about it in a book she won't read.

Why does it hurt not to feel the absence of my uncle? I don't even know what his face looked like. Was he handsome? And what does he have to do with me not feeling now? Why did he end up in this book? In this untitled and purposeless text?

Who am I trying to discover to be? And why is it so hard to be when I don't (re)recognize my own history?

The crisis and the lack

In during my 29-year crisis, I rediscovered tastes I had forgotten.

At 18, my then-girlfriend and my mother pushed me into college. She dreamed of a government job I never wanted, and she wanted—or needed me—to follow in her footsteps. The subject was omnipresent, and I, rebellious, innocent, and raw, exhausted myself. On the other hand, my father said that journalism "wasn't for me" and that I should do the advertising he had started, in which he won awards, but never insisted.

Have you noticed how I use possessive pronouns? As if their desires were also mine, as if we, the three of us, were one thing. And we're not. We continued like this for a few years: I studied abroad and returned, the topics remained the same, and I wanted to escape, to be omnipresent only in my desires.

I got a job at a big production company and stayed in the film industry for ten years.

Ten years.

A decade.

Devoting body, soul, and sweat to a profession I didn't want, thinking

I could escape from myself, not looking in the mirror, and facing the challenges of being who I am. I dropped out of college, learned to produce, budget, and write, took courses, met people, moved out of my house and neighborhood, and traded Santa Monica for Wall Street.

I distanced myself from myself by choice or out of fear. In my four years living in São Paulo, I distanced myself from everything familiar to me, traded newspapers for spreadsheets, and poured out thousands of ideas, plans, needs, and frustrations on bar tables.

I interrupted myself.

Like a child hiding behind the curtains, I pretended to erase my sorrows. You can't pretend your whole life.

I returned to my roots, not quite knowing what to do with my 29 years and no direction, with an almost childlike hope of someone who doesn't know the due date for the gas bill but, deep down, knows. And knowing can either set you free or limit you.

I created conditions to reinvent and rediscover myself. To reconnect with the college student within and say: "Now I'm going!"

And I will do so because I can give myself the support I always needed I learned to value my time and fill my voids.

Today, I am free from my own shortcomings.

I just didn't want to be me
    I wanted to be you
    Not to learn from you
    Or even the other you
    Anyone really
    but me

Imagine how much easier it would be to be that Brazilian journalist who has lived in New York for 500 years.

Or that other one from Brazil, who is super successful and is everywhere I want to be?

Or even that blogger who isn't all that great but has more relevance than me.

Actually, anyone is more relevant than me.

Even the one I judge for starting a podcast "late," after my wave had passed; that other woman with a weird accent; the one who writes without proper grammar; the ugly one; the misfit; the one who doesn't care about judgments; the biggest one in Brazil; the smallest one in the United States.

All of them are more relevant than I am.

I wanted to be anyone, really, anyone but me, and I'll lie to you. I'll say

I want to be anyone else because I like the spotlight. It would be easy if that were all. The spotlight is easy. Anyone fitting the "standard" can find the spotlight, but it's not just that. My ass doesn't fit the "standard," but it's not just that.

I don't want to be me to avoid dealing with my pains, my past, my frustrations, and my trauma.

> I don't want to be me
> because being me means
> exposing all of that
> dealing with all of that
> reviewing all of that.

Being me is the drama overflowing from the blank screen with the blinking indicator. It's having memories that aren't always good. It's having the past very present to use as an excuse not to be anyone else but me.

Numbing

> Numbing myself was the only way to stay alive
> Without feeling
> Without connecting
>
> Running over all emotions to feel them later
> To feel them now
> To cry during a yoga class
>
> On that night

Darker than all other nights before
No hangover was ever so strong

I had given up on myself
After everyone else had given up, too

I could only count on myself
I was too heavy

But I showed up
With difficulties and setbacks
But I showed up
I changed
I felt
I was disappointed
I returned

From the darkness
From nothing
From the darkest blue
From the absence of light
From the absence of love
From the absence of care
From the absence of myself

And you still dared to judge me
After all of that
After not dying, solely and exclusively thanks to me
You thought you could tell me what to do
Nobody who comes back from where I came back listens to someone who's never been there

I came back because I went there to find myself
Alone
Without you
I came back to myself

For myself
In the best way I could
And today, without numbing agents
I know I came back because I went there to find myself.

Thoughts of a Second

Do you know those thoughts that last a quick second and change your whole life?

For a second, and nothing more, you live a new story with conflicts that never appeared and will never be resolved.

Only in that second do you change your name, address, interests, passions, and problems.

You think about how it would be to surrender to your desire for a second and
poof,

everything disappears.

After turning 30...Now, 31

Who am I in the driver's seat?
Not in the literal sense, as I did not pass this test, but in the figurative sense.
Besides being a daughter, niece, wife, and friend, who am I?
Independent of titles, last names, traumas, exes, who am I?
Beyond the chaos of my mother's house, the gypsum walls, the cold strolls, and the giant TVs, who am I??

Who am I in the house that is also mine?
Who am I after the panic attacks, the medications, the therapies, the

care?

Who am I who needs a bucket of coffee in the morning, who doesn't like waking up early, but who also can't stand staying in bed late?

Who am I, besides all the things that are not me but still define me?

There are days...

There are days when we want to go back, go back home, go back to our mother's womb.

There are days when we do everything wrong, and it's difficult to face ourselves.

There are days that drag on when we need to remember that we are not alone and that even if everything seems to be going wrong, facing challenges is already succeeding.

There are days when turning 30 feels heavier, and you just want to wake up at 40 when you'll be more mature and reasonable. There are also days that could last an eternity for you to enjoy every millisecond of that happiness.

There are days when drinking with friends brings joy, and there are days when drinking alone causes harm.

There are days when frustrations loom larger than my one-meter-fifty--four body, and everything explodes.

> On those days
> I just want it to end
> I write to forgive myself
> It's hard to see a future

Reality dims dreams

Self-love dissolves
Lack of courage to face myself
Light invades and disturbs
On those days, I feel like not waking up

Invisible

There are days when the cloak of invisibility hits harder. We look at the world, and it seems like nothing will change, even as we take on more responsibilities, feel our feet firmer on the ground, and trust in the future.

I feel invisible when I interview a badass woman about a crappy time when I read the account of another woman who was also called "honey" and another who was told to "calm down" and also had her pain ignored.

Then, #metoobrasil started, and I realized that I wasn't the only one who gave up on (or postponed?) a dream because of some jerk guys. If it were just me, that would be okay. I could handle it, believing I was invincible, untamed.

But the one who speaks out doesn't! She knows she is vulnerable and, therefore, courageous. So, she faces everything head-on and lays out the carpet for the next one to walk on, even lying on the ground.

Then I can't take it

I feel powerless

Incapable

Invisible

I don't want to follow the destiny they wanted to impose on me or succumb to the curses they tried to push on me, but fighting against what they think is right for me requires more courage than my invisible being could bear. Courage is uncomfortable.

Defeated

Do you know those movies in which the main character's transformation from Cinderella to princess is depicted in fast-paced, dialogue-free sequences? Those types of films in which Anne Hathaway transitions from an unkempt woman to a beautiful princess or important secretary?

I agree. They are modern versions, yet dated, of Cinderella.

So, I play these movies in reverse when I see one of my victories and compare it to others' victories—always better, always shinier.

For me, the passing of time means leaving this princess state where I breathed for a few seconds and returning to Cinderella without a happy ending.

No happy ending.

Carrying the failure of not having the courage to continue, the pattern of always being defeated. Waiting to be saved while I linger at the bottom of the pit. Clinging to the scraps of those who (and how could anyone) love me.

Regret

If regret could kill, we'd all be dead. It's the famous saying (help me out here, Pitty): "Let he who is without sin cast the first stone." And we all have at least a part (I have the entire roof) of a glass roof, maybe even made of crystal.

It is thin, elegant, and precisely placed in the part of the story that hurts, where the wounds linger without bandages or forgiveness.

Not asking for forgiveness is normal, but it gnaws and, one day, floods you. I no longer forgive myself for the many mistakes I've made in life. Who hasn't?

We think we come into this world and won't make any mistakes. We believe we'll shine and give our best and that mistakes won't be repeated after some time. But let me tell you something:

- Life is not linear.

It is in suffering that we learn, not necessarily to be better. We truly learn to fall without hurting ourselves so much.

A Request for Forgiveness

Believe it or not, when I encounter a woman who is in a mental battle from which I have already emerged as a winner, I tend to reject her. Not out of malice, I just don't want to face that fragile and insecure being of my past because maybe my victory is not yet that consistent. Facing my reflection in another body could bring me back to that place.

And it can't. Let me explain in detail, departing from my usual superficial style, but feeling that a deeper dive into the topic might be helpful so that generations of women do not feel as lonely anymore. Pretension? Perhaps, but let's try to relate to this woman writing this book.

First of all, I am not the same woman as three minutes ago. In this second, I carry victories and defeats that are part of my foundation, and there is no reflection that can unsettle me, just as there is no magazine cover that can make me thin or cream that can remove my wrinkles.

Got it?

Let me be clearer. I am part of several women's support groups; I have a podcast with a friend; I am married to a woman; I am in women's work groups, and so on. Still, whenever I encounter a woman who reminds me of a dark moment in my life, I tend to distance myself from her. I forget that my pillars are strong, built with the strength of women who graciously pulled me to where I am today. Perhaps it is still difficult to trust my foundation, which had been shattered for so long.

"But you can trust," I tell myself on hot and agitated afternoons.

– "Trust and ask for forgiveness."

Self-Portrait

I've always seen myself as a helpless person—seriously, without drama—at least for now. What you imagine of a helpless person are subtle features and light traits—a delicate person, and I am everything but delicate. I am a grumpy shorty who rarely laughs! That's me.

I've never found it easy to make friends, never been challenged by internet trends, and never had anyone sit next to me on the bus—which I thought was great in the heat of Rio de Janeiro—and only now do I realize why, as I noticed myself as a frowning person.

My wife says I have a muscular forehead. It must be because I spend half of my time with a frowning face. For me, that's normal. Maybe it wasn't always like this. Perhaps I left my loose smile behind in phases where I needed to protect myself from everything and everyone.

The other day, I got into an unnecessary argument—as most of them do—because I was rude to someone who was just venting. I have this rudeness in my speech; maybe that's why people don't challenge me on the internet. I wouldn't challenge myself, either.

.

The Beginning of the Turnaround

You must have realized that this is not a book with beautiful romance and inspirational stories. Not that there isn't a bit of romance and overcoming challenges present, which in itself can be inspiring, but it's not the essence of my book.

In the first part, I wrote that I didn't want to dwell on introductions. However, I think I need a few paragraphs to share the agony I feel while reviewing a manuscript that reveals so much about what I think of myself, about the distorted images of who I am, and about what constitutes achievement—I still cling too much to what was considered successful in Rio de Janeiro in the 1990s.

I Could Have Died

I always say that I could have died in conversations about my personal growth. For a long time, I dulled myself to avoid experiencing the harsh reality of being the product of a dysfunctional home.

But when I boast about my ability not to die, in reality, I want awards for not having died, for having had strength, for having overcome my darkness as if the greatest reward wasn't my very own life, and as if a trophy could solve my personal issues.

Because I seek reassurances, but who doesn't? A prize for simply existing is the guarantee that I am a person, so I can never again be marginalized as an object that wore out from use, with unruly rebellion.

I am afraid to live and sometimes to die because, on the fringe of family love, there was only the dry sob swallowed on the solitary trip on the passenger's seat and the permanent abandonment when the party ended, and everyone left. We were left alone, just me and her, on drugs, separated by a wall, without rescue, without affection, without salvation.

The Peace I Can Afford

Today, I had my first video therapy session on WhatsApp. They say peace of mind is priceless, but that's not entirely true. It's only priceless for those who already have peace. Those who need and can afford it should seek therapy. I know I feel lighter after reconnecting with my therapist.

When I moved to São Paulo, it took me a while to admit I needed help recovering from recent traumas and the abrupt change. Today, at 30 years old - I'm not sure why it seems so important to mention this age - perhaps I'm afraid of succumbing to the media pressure to pretend that women do not age and forget, without even realizing, that I am aging. I find that more liberating than the fear they try to instill in us.

EFinally, I looked up my Brazilian therapist's phone number last week and sent her a WhatsApp message. I said that I have been living in New York for six months and would like to resume therapy because I have been experiencing a lot of anxiety.

Today, I congratulate myself on the initiative, which was not premature but mature. I like to observe how I've matured as if I were looking from the outside. Observing brings me peace, the kind that I cannot afford.

The First Time I Felt Comfortable

This week, I felt comfortable in my body - for the first time in my life - during a yoga class, and it wasn't with my physical body.

I felt comfortable with my essence, which has finally found itself and sewn itself into the extremities of the body it inhabits.

Every now and then, my soul tries to fly as if freedom were only possible outside, but consciousness, patient as it is, adjusts itself and brings back the need to be whole.

Mirrors Are Hard To Face

The stranger who lives in me greets the stranger who lives in you.

You know what I found out? I can be the auntie from Facebook who doesn't understand anything and asks the most disconnected questions in a classroom, and that's okay. This kind of embarrassment can also happen in a conversation with friends while sharing an opinion about a subject I don't fully understand, and so on.

I always wanted to be cool and avoid my natural instinct of being strange. I wanted to fit in and hang out with fun people. My mother hated my friends and called them "bad influences" - and maybe they were, indeed - not by the concept she had, but because I distanced myself from resembling myself to resemble them.

It's okay not to like the strange person in the room who looks like me - mirrors are hard to face. It's OK to take time to understand something and not be the smartest person in the room, although I've learned that aiming for the opposite is right.

Thankfully! It's tedious to disagree just to be different. It's pretentious. And this is coming from an Aquarius girl.

If I arrive late, I cannot keep up with life as if I had arrived on time.

I arrived late in this city. Thirty years late. In the past, I thought I needed to arrive somewhere, do something until the end, and immediately see results.

Life is not like that.

I felt less pressured. This is good in the United States. I think there's a whole beauty industry here, but ordinary people are less tied to that kind of pressure. At least that's how I feel, but again:

I'm already 30 years old.

I'm shy and used this mask pretending to be cool to fit in a world I didn't want to be part of—a place where I felt oppressed.

This got confused, right?

Life is not linear, nor are thoughts.

The role of a writer is to give meaning to the confusion of the soul. I think I want to say that it's okay to be strange. And for the first time in my life, I feel comfortable with that.

The Advice We Listen To

We take little or no care with the advice we take in life, right? It's not because someone is giving a speech on a random subject or is a boss that they are clear about that subject.

Let me explain: In my podcast, I've been asking my interviewees about the best and worst advice they have received in their careers, and the things I've heard are hard to believe. In addition to the work environment, we also hear advice from relatives, friends, and authority figures and forget that much of the advice simply doesn't fit into our lives. It takes self-awareness to listen to advice or look up to someone.

I Want To Live Only In My Dreams

Once, I applied for a job at the UN. Despite thinking the UN is a respectable organization, I never considered working there. Truth be told, I never quite understood what they do. I know they are a respected institution that brings world leaders together to discuss solutions that I never see in newspaper headlines.

Forgive me, UN, but I never wanted to work there. When I applied for the job, I was living in New York for a year, and for the first time in my job search, I applied for a position at the UN.

Imagine how chic that would be. My mother could boast back in Brazil that her daughter works at the UN. Other people probably don't value the UN that much, but my mother does—and so do I. Imagine the hassle of being a babysitter for world leaders who don't care about human rights. Imagine spending your entire career trying to persuade people in suits to care about preserving the Amazon instead of cosmetics. And without letting it turn into pasture for cattle. Mission impossible.

I "don't give a damn" about world leaders, except for Queen Elizabeth II, who don't call the shots since 2019 and shouldn't be there for so long, but I find her cute. I love envying Kate Middleton's height and elegance, but I love the transformations Meghan brings to England's famous and privileged family even more.

In Brazilian Portuguese, instead of saying "royal," we say the "real" family (A Família Real). So, I refuse to write "real" family in Portuguese. What family isn't real? A single mother struggling to raise her children is a real family, just like my wife and I and our plants. A very real family indeed – complete, with officially signed papers and all. We are a newly recognized real family, and we should celebrate that.

I write so that one day we won't have to celebrate this victory anymore, so that one day we'll only say "marriage" and celebrate love.

And excuse me, UN, but I prefer to write for real people in a clumsy style that I'm still trying to figure out, with curse words, slang, and little punctuation. I don't give a damn about men in suits and ties because:
1. I don't even like men that much.
2. I prefer people with broad smiles, on the beach, talking about life.

It's not that the beach crowd doesn't know anything about economics We know very well when the price of tomatoes goes up, even if we're not "beautiful, proper, and homely" as they say in Brazil—bela, recatada e do lar. We know about economics because food weighs heavily on the budget of those who leave home at 4:00 in the morning, don't shake hands with any executives all day long, and return at the end of the day hoping for a better life.

The UN is not my dream. It never was because I prefer people with souls. To me, standards sound like another way to stiffen us in a society that wants to suck out the best in us: our authenticity. I almost fell into the trap of living a dream that wasn't mine for the status it would give me. And let it be clear that I have nothing against the UN or against those who sweat to change the deepest ingrained values of a sick society. The point is that I didn't want to live a dream that wasn't mine.

I also don't want to sell my art at the beach boardwalk. I want to leave it here, available for anyone who wants it. I want to have a livelihood that gives me freedom and launch books that my friends will buy "to show support," as they are the only ones who listen to my rants at the bar tables.

That's what I want, even if I don't sell one book, which is unlikely because they know I have a list of people I "curse" when they don't support my work. And if they didn't know about it before, they know now.

Rooted In Me

What in me wants to buy what I don't want?

That trades reading a book for online shopping out of fear of "being too out of touch with the world?"

Within myself, it is inadmissible to be even stranger.

Then, I tried to fit in

by buying things I didn't need,

swapping books for sneakers

just to not give in

not to surrender

to be more mundane

to continue propagating the standard that my friend, at 10 years old, pointed out:

- You are the most unique person I know, trying so hard to be the most normal.

    I was.
    I am.

    Why? I don't know why.
    The unique one is not loved.
    I want to be loved
    to fit into some standard
    to have a place
    to feel a little connected.

    Being unique is rooted in oneself and requires work
    I didn't want it to be work
    I wanted to be normal
    It gave me more trouble.

    Damn.
    Will it pass?

The Spirals Of Life

This part will be a little more mystical, but I swear it will be worth it. Although I consider myself the strangest of creatures and try, with all my might, to hide the potential of my 5'1" height, there are subjects that I cannot help myself and need to touch upon.

The Spirals Of Life

I had an upbringing that, in today's terms, we can call holistic. At home, no one was extremely Catholic, Buddhist, or spiritual, but everyone believed in a little bit of everything.

In childhood and adolescence, I heard and read about the mysteries of the universe, psychology, behavior patterns, neurolinguistic programming, astronomy, astrology, and basically everything related to the sky, the water, and the air. Books like The Secret and Man and his Symbols permeate casual conversations with the intellectuals who brought me into the world, and science and beliefs intertwine without ceremony. I grew up believing in a little bit of everything, and I still do.

Recently, I went through a coaching session, and in one of the sessions, I heard a sentence that blew my mind:

Our life is made of spirals.

According to her studies, we repeat patterns every seven years, and thinking about this spiral, instead of imprisoning myself in fear of the next wave of mistakes, liberated me.

I still find this new sensation of life's infinity strange. I would love to notarize, with a recognized signature, a certificate of overcoming trauma, but it seems that this certificate won't happen anytime soon.

They say that enjoying the journey is the secret to dealing with anxiety. I'm still testing it.

P.S.: It's not about being conformist; it's about preparing not to fall so hard in the next blow.

Warning: Long Text

In the past, I thought that following a religion was a sign of weakness. Today, I know how foolish I was and see how much Buddhism has changed my life.

Taking responsibility for looking inward and causing your own human revolution takes a lot of strength. I am grateful to everyone who has crossed my path and, in some way, influenced my decision. Especially my father, who always gave me an expanded view of spirituality. And, of course, my cousin, who literally led me by the hand.

On August 28, 2016, I officially became a "little Buddha." Gratitude is all I have in my heart, and I cannot express how happy I am about it.

The People in My House

Today, I decided to organize my virtual paperwork. Since the beginning of the year, I've been trying to interpret the message the universe keeps sending me. I listen to a weekly astrology podcast that repeats the following message:

"Aquarians need to tidy up and get organized."

As I don't usually stir up trouble with the universe, here I am, since the beginning of the year, organizing drawers that keep getting messy, ideas that persist in confusing me, and documents that continue to intertwine like cassette tapes. I don't even know where to look anymore, but I always find a mess to clean up.

Today, my wife said she will organize my magazines. I have never subscribed to a magazine but occasionally spend hours at the newsstand, choosing some to take home. Not long ago, a People magazine arrived in the mailbox addressed to me. I never subscribed to a gossip magazine, but it arrived in my name, and they kept coming. I decided to keep some to create a dream board. They say it's good to visualize our desires coming true. It's all part of *The Secret*, you know?

The problem is that the *People* magazine and I have nothing in common, and even though I want to be special, I have a strange taste for the ordinary. I even tried to adapt to the dream of living in huge houses, attending noisy parties, aiming for senseless awards, and wearing tight dresses, uncomfortable shoes, shiny accessories, and matching bags. It didn't work. I fear big houses, get tired of the noise, and have an aversion to tight, fashionable, and uncomfortable clothes. I won't be hypocritical in questioning the awards, as I also enjoy giving gifts to my friends. What exhausts me is the endless ceremony that combines everything that drains me.

I don't even go on social media on Oscar Day. I just like to look at the dresses I wouldn't want to wear. In fact, I need to reassess that. The magazine I received didn't have a photo of someone practicing yoga, reading a book, or enjoying a cuddle on the couch. Come on, People! How am I supposed to visualize my future if you don't publish a lesbian couple sitting in front of the fireplace? Is it forbidden to publish laziness in *People*?

Limits Are Necessary

If you don't establish the power others may or may not have in your life, you will always be at the mercy of their desires. Setting boundaries means living your life according to what works for you.

For many years, I lacked affection, self-love, and self-confidence. I don't even know how I got to where I am now. I swear. I had very enlightened people in my life who treated me with respect and love when I couldn't even do it for myself.

But not everyone was like that—quite the opposite. Several took advantage of my lack of boundaries to benefit from my love and generosity.

I wasn't wrong to be generous. It was the people who wanted to suck out all the best in me and leave me behind after all I had was gone who was wrong. Unfortunately, those people will always exist. Maybe I was that person in someone else's life - who knows? We need to learn to set boundaries, and if your boundaries bother someone else, that's their problem.

It's your commitment to yourself to know how far to allow something that either benefits or harms you.

I'll tell you a secret: no one comes looking for you at the bottom of the well. Getting out of it is a lonely path.

Be careful allowing someone to throw you there.

The Universe Listens to Our Intentions

We are experiencing what should be one of the most challenging years of our existence. Notice that I wrote "challenging" and not "difficult."

I have been reflecting for some time now on the effect of our intentions in practical life. While it is expected to ask for things in prayers, we rarely pay attention to the intentions we emanate outside of that moment of deep concentration.

Have you ever thought that God, the Universe, and even our minds were not educated in the language of humans? When we ask for something, we never think about the receiver, and it seems to me that the receiver has a much more subtle language than that which we can carefully formulate in prayer.

The Arrogance of Writing About Life

I spent years trying to figure out what in me is me. Look at the title of one of the chronicles, look at all the texts read so far, my echoes, and look at all my failed professional attempts to avoid doing this: launching a book.

But here I am, right? And now what? What will I do with this dull knife and this old semi-cured cheese, forgotten in the fridge? Maybe I should swap the knife for a grater. What's the name of that MasterChef host again? I need to make sure I can make this swap. Damn it! Time's up, and I need to plate up.

So, here it goes…

Blank Screen

A blank screen. That's how great love stories begin. It doesn't take great artists to portray true love stories. They arise naturally, like a flower blooming in spring.

Guiding Questions

Trying to find a path, I discovered that my answers were in the questions themselves.

When I moved to New York in 2018, I wanted to change the course of my career. I intended to write about everything, but soon enough, I understood that I needed to understand myself first. But knowing oneself requires a lot of courage, and we are not always prepared.

As I tried to find myself in my career, I asked questions and met incredible women who are also going through or have gone through the same things I have.

What Did I Want Them To Find?

I have a spreadsheet where I jot down topics to write about here, in my newsletter, or wherever appropriate because an artist is not that group of people who sell trinkets on the beach. It can be, but an artist needs a mental organization like everyone else; otherwise, nothing comes out. We need many things, including mental organization, not only mental organization. I organize my mind in spreadsheets. Looking at my creative planning in Excel cells gives me much inner peace to create, edit, and continue.

One of the ideas on the spreadsheet is titled: I was never found, and now that I'm gaining relevance on social media, I wonder what I wanted them to discover exactly.

There is an illusion that things are easy, that a powerful agent discovers people, and that life is all figured out.

Spoiler Alert: it's not!

Several of my guests on the podcast have already discussed this, and I am increasingly convinced that there is no easy path. We need to carve out our spaces. The challenge for someone else may seem easy to you, but for them, it's not.

Contrary to what I was taught, showing your work is essential for people to

get to know you and recognize you for what you do.

I Need to Address the Lack of Credibility I Have With Myself

I have been in quarantine for 23 days. I remember the last time I went out; it felt like last year. I went out to support a friend at her blog's third-anniversary celebration event. I miss going out, even though I didn't go out much, on average, compared to others and even to myself at other times.

But how I miss seeing the streets, being in open spaces, the warm summer afternoons when I wore denim shorts, feeling the fresh smell of things, and even the dirty New York subway.

Summer—that was last year or two years ago—time froze. And I worried about whether there will be summer again. Not the summer itself, which never fails to arrive, but what I meant was: Would we have the chance to go out and enjoy it?

Stay home! Stay home! Stay home was the city echo disturbing my peace.

It is spring in New York. From the window of my house, I see a wide avenue with few cars on the street. Most of the week, I open the window and feel a pleasant breeze below 70 degrees. The days pass slowly, and the soundtrack of sirens, unbearably loud, disrupted the silence of that ghostly city.

I limit my outings to the grocery store across the street and spend the day refreshing the Amazon page to see if I can get deliveries of basic supplies to avoid going out. I even found the products but couldn't find a delivery slot. The life of a privileged quarantiner is tiring.

Back to my window: I see the flowers growing on the avenue, I hear the loud sirens, I think about how strange it is to see the streets without cars, I look beyond for an answer, and I pray to the wind.

I think of all those who cannot contemplate life. The beers I bought that should have lasted at least two weeks are gone within three days. I take a yoga class, call friends, join conferences with people I don't know, break my own rules, take my vitamins, roll in bed, and wake up.

It all started again.

When I think of doing something different, I conclude that tomorrow is a good day to start, but I need to address this—not the lack of initiative but my lack of credibility with myself.

Therefore, I start something new almost every day. If I didn't, there would be a beer shortage in the market. Starting something new is stronger than me; I just find it difficult to continue like this text, which I've been writing for hours and don't know how to end.

Envy Is For Those Who Can't

I don't know if I'm lazy about Sandy (a famous Brazilian singer) because I find her incredibly dull - like my cousin - or if I'm envious of my cousin because I find her incredibly perfect - like Sandy.

By the way, I think they're both the same age. How can you not envy someone friendly, successful, intelligent, talented, and beautiful? Not for nothing, cousin, but you've set the bar high in the family. You made it really hard. Do you think my father wouldn't want to say that, at 30, I already had my own apartment, a nice car, a light-eyed child, and still had the energy to travel with a newborn? Of course, he would. Just like I would love to go around saying that my father has a boat in Bahia, and at this moment, he is crossing the open sea like my godfather.

I would like to, but not even my father—who wanted me to be successful by 30—thinks it's cool to work for a cigarette company. I don't think it's cool for my father to venture out at sea without a responsible person by his side. I continue not to be successful, and my father continues to live in the middle of nowhere. We have different ambitions, and in silence, we believe we are happier.

I Would Be Great At Everything

Because I never believed I could make a living from writing - it's still hard to believe, actually - I sought direction from other lives on how to lead mine. And when you're daydreaming, it seems like you have the ability to do almost anything except for what you need to accept that you must do. And absolutely any solution seems easier than carrying the heavy burden of being who you are.

I didn't want to live from writing because I didn't want to expose myself. Writing is putting your deepest experiences into connected (or not) words. I was foolish because to accomplish any task, you have to expose yourself, show your vulnerabilities, admit you don't know, ask questions, and be ready to make changes.

I wasn't ready to make changes. Changing means being secure, and I, who barely stepped on the ground, couldn't edit my life without sinking even deeper into the mud.

The "would be" never "is." And I needed to be.

P.S.: Be bad at everything else—be terrible! That way, it will be easier to do what you love.

She is the one who trusts, not me

Recently, I received an email that filled my heart with joy, doubts, anxieties, and accomplishments—all at once, without boundaries or requests for permission. I explain why: I am a good person (great, I would say) in facing my superficial fears and putting my face out there without overthinking about the next steps needed to execute the crazy ideas I have every second.

Last year, after a very difficult process of treating anxiety, I started attending all events related to women and writing. At one of them, I said I was going to start a podcast in English to talk about women and the job market in New York; at another, I introduced myself to a writer and said I would love to interview her.

Cut to the scene. Months later, here I am with a book I received from a woman who trusted me before I even trusted myself.

This podcast I wanted to launch already had an edited and thoroughly mixed episode that was not uploaded at that time due to a lack of courage (now, it has two uploaded episodes). Every time I edited it, I found my speaking bad, my questions silly, and my English terrible.

The author of the book I received doesn't know, but I have re-recorded (countless times) my lines in the pilot of the shelved podcast. She doesn't know I have only shown this podcast to one person. She doesn't know that I'm terrified of not only publishing that one but all the other projects I keep hidden away in a box labeled "insecurity."

I lean on and drive myself in the strength of incredible and courageous women with whom I made sure to surround myself.

I wanted to say that I am living the dream of receiving a book that I will read to create a podcast. The truth is that the package arrived filled with pride and imposter syndrome entrenched all over the cardboard.

I could even use the quarantine to hide or postpone the interview, but the internet... Ahhh, the internet! It doesn't catch the coronavirus, and you can record a podcast without leaving your home, right?

Oh well, the internet! Couldn't you give me some slack on that one, too?

Oh, 2020! You could not even be like the other years and distract me with thousands of projects I'll never put out there.

For today, that's it, but don't be fooled. I've procrastinated more than I should have. I read everything that wasn't that book, played all the video games I could until I got bored, wrote two projects I'm trying to produce, used the excuse of rebranding, and, finally, I knew I needed to get the podcast out there by the end of that week because I'd already talked about it - and I'm still talking about it here, *really*.

I Don't Need to Defend Myself Anymore

Receiving feedback, including from this book, is no longer a reason for me to hide in a cave, afraid of getting scolded. There are no scoldings in adulthood. There is only you, fighting for what is right, raising your hand, asking, and asking to be heard.

I no longer need to hide to defend my self-esteem from others because I am no longer trapped in their opinions about me.

Defense is a mechanism threatened animals use, but I am no longer threatened. I am in the process of evolution, but of something formed, something that already is, that adapts when necessary, but that does not need special authorization to be, and therefore can never be threatened.

The Arrogance Of Writing About Life

I wanted to be a star in the circus where my father used to take me as a child: climb on stage, stand out more than the clown, and be celebrated.

I don't know where this desire to stand out comes from, and I don't even know if I can justify it with psychoanalysis. It would be great not to admit that I really like being praised and blame it on my mother, who loved me in her Capricornian way, or even say that it's a result of the abusive relationships I've lived through.

Freud, this one's on you, buddy. As a writer, I need to be blasé. I don't like to stand out, observe more than speak, look at everyone from a higher point, and boast about my pretentious intellectual capacity as someone who likes (and even knows a bit of) playing with words.

Writing is playing at having control of destiny, being somewhat in charge of oneself, of the characters, the end of the story, and having objectives. It is hard to admit to oneself what one wants. I blame this condition of daydreaming on astrology.

We, who write because we feel compelled to, were born with needs just like everyone else. We are admired yet ashamed to live with such dissonant traits. Then, we look at everyone from above, but not out of spite.

Something happens inside when we put on our writer's face, and it's hard to realize the arrogance of being.

Just Because You've Never Done something Before Doesn't Mean You Can't Do It Now

Often, I'm afraid to do something just because this is the first time I've done it.

I live in the mode of a thousand "ways to avoid new things," which is the complete opposite of the "a thousand things to do before I die," highly valued by our society. My modus operandi protects me from frustrations and traumas but also prevents me from living fully.

I've always written. Always. Since I can remember. My first typed fiction story was written when I was seven years old.

I immediately became interested when the Love Notes to NYC project was launched. The project, an Instagram account that was also supposed to become a book, brought together love notes to the city during the 2020 pandemic. I supported it, as I thought the initiative was great for everyone, not for me, never for me. I couldn't even write a single sentence in English, let alone a note to honor this erupting volcano I call home. I was invited to participate in the project, and I didn't write only a sentence, I wrote an entire poem because I felt empowered by the trust placed in me.

Of course, I was afraid. After spending decades writing just for myself, little by little, I started showing some of my creations—in Portuguese—to very few people, and rarely did I show the most intimate text to the public.

I changed my "bio" several times on the day I identified myself as a writer for the first time, but I did change it.

Later on, a poem I wrote - in English - was published.

I made it here.

It is important to point this out here because there seems to be so far away.

Larissa Rinaldi

I Really Need To Stop Being This Person

I can no longer have the folder of unpublished texts fuller than those of published ones. Not for others or even for myself. I just need to stop being someone who hides in need of guarantees and craves praise. Maybe for myself, I need to stop being this person. I need to publish everything I want without caring about what others will say - others speak without fear and the courage to be. It's the role of others to say what they are not. It's my role to find myself amidst fears, whispers, desires, envy, and rebellions.

I need to detach from the childish instinct of wanting to protect myself from feeling. It is not the haters that I need to protect myself from; they only see faults, but nor is it those who love me because they focus on my qualities, even if I make the most naive mistakes.

I need to publish myself, even without deadlines and contracts. I need to stop ignoring my existence and unleash the voices that shout louder than fear, inertia, and rejection.

I need to be published, author Lari Rinaldi. I am going through a crisis with Larissa, and it's a bit too pompous of a name. I need to publish myself and not care what others say. I need to publish myself and listen to those who already believe in me, including the newest member, Lari(ssa?) Rinaldi.

Texts have immense value to authors, or they don't. I have read many books that have saved me, but even if they did, who in the world is more important than what I need to be for myself?

I need to protect myself from whom? Those who love me love me with my faults; those who hate me do it despite my qualities.

The fact is that I can no longer live without feeling the pains from which I try to protect "my" little being from my own self.

```
Your Name    Larissa Rinaldi

Pay to the order of    My Dreams              1,000,000
  One million pages in a drawer

                                    Ms. I Need Guarantees
```

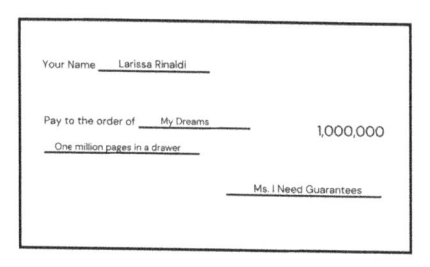

No One Doubts a Confident Person

Those who work publishing content on the voracious social media platforms need support, especially in the beginning. That's when we turn to those closest to us, people we like and respect, the well-known friends and family.

These people are not always ready to offer support, and the reasons are numerous. But you, a small content creator, might become demotivated and sad because you don't see your work yielding results.

It took me almost years to start appreciating the results of mine, but that's a story for another day. Today, I want to mention a post attributed to Viola Davis:

"Sister, I don't care if you get zero likes. Keep promoting your business. They are watching. Trust me!"

Even though it's difficult to continue without the support of those you admire, you must keep going. You must push forward even when you don't feel like changing out of your pajamas, inspiration is lacking, and procrastination sinks you into the couch.

Beyond the numbers, what grows when you promote your business is your confidence, and no one doubts a confident person.

I Made It Here

I'm living a dream, and I feel more confident every day.

I want to share a story for those who are insecure about starting something new or launching a podcast. When I arrived in New York, I went through a lot of anxiety processes. The only thing I knew was that I finally wanted to work with content, but I had no idea how I would do that.

At the end of 2018, I created a podcast in Portuguese called Tudo Sobre Coisa Nenhuma (All About Nothing) with a friend who was also my co-host. At the beginning of the pandemic, she needed to take new directions, and I was left alone running the project. I started inviting women from my circle to interview them, and now I receive guests I've never seen in my life.

In December 2019, I thought about starting a podcast in English to build my portfolio. Not so long after, I recorded the third episode, interviewing the author of a recently read book. She was an executive at Elle magazine. Can you imagine my nerves?

This week, I also interviewed some influential Brazilian women in Portuguese. The interviews in Portuguese help me with the ones in English, and vice versa. I'm sharing all of this because I used to think of myself as "the horse poop," and having such badass women accept being interviewed by me no longer makes me feel like a total zero. I'm living a dream and feel more confident I'm doing a good job.

If you're feeling insecure, show your work to badass people who will push you forward. I had and continue to have this support, which brought me here.

How Do I Go Back?

I was wrong. I didn't want to find myself. How absurd is admitting you're lost to want to find yourself? Where did I get that from?

Like, now, instead of finishing this book, I just want to be playing video games, looking for my next vacation destination, sleeping by the air conditioning, asking my psychiatrist to let me go back on medication, crying in my wife's arms, reading my Facebook and Instagram feeds, scheduling a manicure, cursing at those who answer my questions with questions, watching Big Bang Theory (with Friends on the other screen), reading a book, eating ice cream with syrup straight from the container, forgetting I received compliments, ignoring that I was born for this.

– "No one is born for this!" I wrote the other day.

I can't be contradictory, but I already am changing my mind: "

– "If I wasn't born for this, what was I born for?

How to get out of the crisis?

How to finish the book?

How to launch it into the world?

How to live with your ideas living in other minds?

How to tell the publisher?

How do other writers do it?

How to make money?

How to go back to procrastinating?

In another tab: tickets to Universal Orlando.

It's Better To Give Up

It would be easier to give up. Think about it: it's been three months without actively working on it. I've only had meetings, gathered opinions, and written some stuff (as the poets say).

It's almost ready, but why launch it?

How agonizing it is to put artistic creation out into the world!

It's easier to give up. It's only seven years of work, compiled in nine months, while pouring out – and still pouring out – the agonies of a pandemic. Nothing too major.

It's easier to give up.

No one needs this book in the world.

Not even me.

Not even me?

Maybe I do need it. But I don't "matter."

It's easier to give up.

It already has a copyright.

So what? Why not give up?

No one needs another book in the world.

Except me.

I need my book in the world.

But can I still give up?

Captions for a post – Instagram

Stripping away the masks I use to hide my insecurities isn't something I do every day.

I felt like the worst of beings, and with some help, I'm creating quality content that I'm very proud of.

End of a Cycle

I haven't always been the best person, but I've always tried to be the best person I could be. Life isn't easy; it doesn't always flow, and our decisions aren't always sensible. Over the past ten years, I've gone through various trials, and I wouldn't be here, in this moment of love with the universe, if I hadn't encountered you along my way.

It's Not That I Didn't Want To

    I wanted to.
    I wanted it a lot.
    But I couldn't.
    The pressure was high.
    I set it.
    I wanted to see the end before the starting shot.
    Now I can.
    Now I've arrived.
    And maybe I can actually make it.
    Or maybe I'll just be
    Be what's even more sublime.

How Chic Is It To Be Sublime?

Just the word sublime is already sublime in itself. Sublime is so fancy that if it were a person, it would be Fernanda Young (a Brazilian writer, screenwriter, television presenter, model, and actress). Sublime in tattoos, ironic remarks, and self-confidence.

How chic is it to be more than one appears to be? It is a classy thing; those who have it because they are, not because they have it. Like the residents of Leblon, a Brazilian neighborhood in Rio, with their shelves of read books.

At the beginning of the COVID-19 pandemic, second-hand bookstores were selling books to people who wanted to appear smart on video call cameras—people who wanted to be chic but weren't.

I almost forgot to talk about myself, who happens to be the guiding thread of this book. It's just by coincidence. We didn't sign any agreement beforehand, neither did the book nor I; it just happened to be that way. It exists because I am myself, even without being sublime.

In the past, when I didn't like who I was, I would think about having dozens of purchased books, but I don't think I would actually do that because I didn't like spending money. And how much would it cost me to seem like someone I wasn't? Not that I didn't have any books to show. Self-help on the shelf doesn't look good, does it? It's much better to read Ernest Hemingway, Charlotte Bronte, and a few other names that I can't remember now because my bookshelf is contemporary, and being contemporary is not sublime unless you're Fernanda Young (a late Brazilian writer. Think of Glennon Doyle for Americans), who unfortunately is no longer contemporary either.

What I mean is that being sublime isn't a thing for someone from the suburbs, from Queens or Staten Island, even less sublime or more hipster, depending on your point of view. Clarice Falcão is from Leblon in Rio; Larissa Rinaldi is from Itanhangá. It may or may not rhyme, but it doesn't quite fit in, you know?

Before getting along well with my lack of sublimity—yes, that word exists—, I would think about buying books that I hadn't read to seem like someone I wasn't and impress people who wouldn't make room for me in a room, even if it seemed like I had read many books. Space is about being, and those who conquer it are sublime.

Skid Marks of Me on the Pavement

A *carioca* from Rio

It's ironic to think that my story with Rio began to be written after I left there. Not that the 24 years I lived in my hometown were wasted. I still carry something of Rio within me.

I remember the first time I looked at Rio for what it is: a beautiful city. Not just beautiful, wonderful. Rio is everything you see in the pictures and much more. I didn't live close to the tourist spots, yet I saw the city's natural wonders every day on the way to school.

Not uncommonly, I allowed the city's beauty to invade my daydreams, and I would lose myself in the story I was trying to create. Rio is like that: a subtle invasion to the sound of bossa nova, with a smile on your face and an invitation to samba. Before you realize it, passion has taken over your body, and it's too late. Now, you have to deal with the traffic around the beautiful lagoon between the ocean and the mountains, the politicians in jail, the neglect of security, the parallel power, the scorching beach, and the lack of opportunities.

I never surfed

I tried. Once. I bought a pink thermal suit to go into the cold sea of Barra da Tijuca. So, I went with my mom to crash my then-boyfriend's class. It was a disaster. I was so afraid of making a fool of myself that I cried and made an even bigger fool of myself.

The rest is history. The rest is that surfboard I never got up on, the three times I almost drowned at the beach, the years when I didn't feel like a true carioca because I used to wake up late and wanted to stay by the air conditioning.

The rest might be a chapter about when I lived near the beach and spent hours on the sand drinking beer under the umbrella, talking about music with my friends, and leaning on their shirts from famous bands.

The last one to leave

I'm always the last one to want to leave. Even if the party is boring and I'm falling asleep standing up. But it only happens when I like who I'm with. Otherwise, I'll slip away unnoticed. And I did that in Rio repeatedly. I would go to parties with friends and disappear into the dark streets of Copacabana in the hot early morning. Teenage irresponsibility. Child's luck. No, nothing ever happened to me.

Now, I don't rely on my guardian angel so much anymore; after all, rumor has it they're busy with drunks, tourists, and Pisces people.

Rio is no longer mine

Abandoned for many years, Rio returned to the attention of Brazilian politics with the proximity of the 2014 Olympic games. This was advantageous for the city, which received improvements, and also for tourists.

And the *carioca*? They couldn't imagine that improving the city could hit their pocket so hard. Soon, the playboys from the South Zone will be forced to take a bus to the North Zone and visit its beaches only on weekends. Not that the North Zone is bad, but the upper class of the South Zone can no longer bear the prices in their neighborhoods, just like the middle class of the North Zone and so on. The improvements have come with a lot of real estate speculation, expensive standards, and kiosks that only accept dollars.

It's true, Rio has always been a tourist city, with beautiful beaches and stunning views, but loving Rio comes at a high price. More than once classified as the most expensive city in the Americas, Rio is becoming less and less for the cariocas. The surfer, who applauds the sunset at *Arpoador* and crosses the still-warm asphalt barefoot to get home, already shows concerns never seen before in his demeanor. Life is very expensive. Not only for him but also for the *Girl from Ipanema*, who now only goes to the beach on her days off and is bitter about the price charged for the umbrella (an outrageous $30). Even those who pay in dollars complain about the cost in Rio.

Let alone being classified as the most expensive city in the Americas, Rio de Janeiro - and especially the cariocas - endure the shocking comparison between the value of a home in Rio and Paris. Nothing escapes the overvaluation in Rio: from the coconut on the beach to the daily rate at the hotel. Everything with a price tag is expensive.

The radiant smiles, once indifferent to life's troubles amidst beaches, dive bars, restaurants, and nightclubs, are now bare a yellowed hue of faded and old news. Now, the carioca with a good job and considered upper class in Brazil would take at least 30 years to buy a two-bedroom apartment in the South Zone.

It's not that the *carioca* doesn't like to work, but the traditional hours don't work for him anymore. The surfer needs to take a dip before changing his trunks for a suit, while the *Girl from Ipanema* goes to have breakfast at *Jardim Botanico* with her friends before going to the office.

This is life in Rio: we value and create enjoyable moments daily.

Tourist at home

Whenever I arrive in Rio, I am certain that I will receive a tight hug, eat at a restaurant where I know the entire menu, sit anywhere, and have a breathtaking 360-degree view.

Being a tourist in your own city means enjoying, with a nostalgic filter, everything it has to offer and not worrying so much about the problems that, after all, are no longer yours. Visiting Rio while living in São Paulo meant arriving at Santos Dumont Airport excited and taking off in tears. Why did I decide to move?

– Ahhh, yes! Rio didn't have a place for me.

Jealousy is the word, right?

Of course, I am sad about moving. In 2014, I left Rio, where I was born and raised, to live in São Paulo.

At the time, I didn't think I was abandoning my life. I figured I needed to go and start fresh, and still, I cried for several weeks. I didn't even know the reason. I just felt sadness.

And, of course, when I moved to New York in 2018, I was sad again. In the last few years, my life has changed: I got married, I no longer feel so alone, and

I have a better understanding of who I am and what I want.

Nobody truly wants to leave their country, the place they call home, the people who speak their language, and the warmth of relationships that have existed for years.

I am sad about the move. I wish I had the same opportunities in Brazil, where I was born, and understand the culture. It's not a move for survival. We are seeking career growth, better salaries, and comfort.

I won't be able to work for a few months. We are confident that everything will work out, but I feel sad when I remember that, if I'm lucky, I'll see my old friends once a year, I'll leave my mom, and I think that I won't be able to spend as much time with my dad as I would like. I'm sad knowing it's not just a 50-minute flight that separates me from the Christ statue and the 100-degree heat on the balcony.

I don't recognize the streets of São Paulo as welcoming, and there is no fault in that. The streets of the capital don't have the sea breeze, a juice bar on every corner, an intense heat, and the horizon at the end of the street. It's a different city, and it always will be. Despite having incredible opportunities in São Paulo, I always missed using the beach as a reference for navigation: one way is downtown, and the other is Barra Beach. It's as simple as that.

New York would be worse. I never got over the jealousy of leaving Rio because I had to, imagine leaving Brazil. I'll meet Americans who probably won't like me, and I won't have long-time friends to have a beer with at random bars.

Moving to a new country is like entering a casa do Big Brother Brasil, with no exit date and no prize of 1 million reais (about 250k). It's a gamble. You are against yourself trying to explore the world, searching for something you don't know what it is nor haven't found it yet.

I demand to like myself less

I miss having a friend to go to the mall, window shop, talk nonsense, just hang out.

It's been six years since I left my hometown (ahhh, Rio de Janeiro). There I had companionship for everything I wanted to do. In the last years living in São Paulo, I got used to not having friends and being alone. But that era ended.

I demand liking my own company less.

Real *Carioca*

After living in New York for a year, it was a bit uncomfortable to meet and talk with what I'll call a Rio native - a carioca - for over an hour. Not because he was sly, far from it. He was the type who was raised in the South Zone of Rio de Janeiro, boasting about the past, praising other people's money, and deep down feeling betrayed because Rio wasn't the "Wonderful City" (Cidade Maravilhosa, as Brazilians say) for him.

The cultural shock, which only happened because I left Rio almost ten years ago, came in the middle of our meeting. *Cariocas* have a culture of valuing the decayed, and that's something I can't stand anymore.

Carioca in SP

I have traveled to São Paulo every two years, as far as I can remember, and I have always admired the São Paulo residents' approach to life.

Eventually, I discovered that my issue wasn't exactly with all cariocas, nor did I admire all Paulistas. The fact is that I loved driving to Paulista Avenue, seeing the lights, the absence of a beach, the calmness of holidays, and the hustle and bustle of the week.

São Paulo is a big city like the dreams of yore, with the ambition to live a more fulfilling life away from what hurt me, which wasn't the sea air but had the smell of Ipanema.

Two months in São Paulo

Exactly two months ago, I arrived in São Paulo with suitcases as big as my desire to become independent. The city overwhelmed me as soon as I got to the bus station. It was difficult. I felt suffocated. São Paulo is suffocating. Pressuring. Not getting back on the same bus I came on was my biggest act of courage. I wanted to give up every second of every day during the first few weeks, but I persevered.

– Let's go! I dare you to beat me.

Then, you discover you are capable, and surprisingly, you conquer São Paulo back. Everything is challenging. You are tested all the time. And even a weak person becomes stronger.

Less attentive eyes don't see the beauty of the gray-flooded alleys. Some people don't like it. They criticize. They say it's ugly. Not me. I find it beautiful and intriguing.

Bring on a thousand more challenges, even if they hurt a lot. Today, I'm sure I can handle each one.

Thank you, São Paulo.

Three months in São Paulo

Life comes and destabilizes everything. Thankfully, there is the "comeback." Three months, a lot of pain, some victories, and one certainty: I came here to win, and nothing and no one will stop me.

I am stronger and weaker, more practical and sensitive, with more doubts and certainties. Each challenge is just a push. Let's go, São Paulo. We are together until it is time to say goodbye.

The book and the delay

She was running late, and I knew that if the bus had arrived on time, I would be there five minutes after our agreed-upon meeting time—five minutes isn't really late. But the bus didn't come, and I stood there, engrossed in my book, reluctant to go back home alone.

Whenever I read short stories, I felt like I could write one, too. All I needed were paper and a pen for the words to flow as poetry, a short story, or prose—at least, that's what I thought.

The weight of the two bags I was carrying was bothersome, but I was happy on that gray Monday. I was grateful for doing what I loved, even standing there and waiting for the bus. I observed closely, hoping to see someone I knew. There's that silly habit writers have of thinking there's a story on every corner.

Lost in my writing game, I'm now five minutes late, resting between bags and poles, counting the change in my wallet.

Many chapters later, the message arrived on the cell phone, bringing my thoughts back to reality:

– I swear I'm arriving now.

There wasn't even time to put on makeup. Too bad the book is over. Will I be able to work?

Sunday suffocates you

Sunday suffocates you. Even when it tries to please you, Sunday suffocates you.

It's not precisely Sunday's fault. With the sunlight streaming through the window comes the shadow of Monday. Suddenly, you remember the bills you have to pay, the dog's bath, and the concert you want to attend, but the commitments won't allow you just to stay there.

And Sunday goes on suffocating you.

Unproductive idleness agonizes the body and crushes the heart. And despite the lottery results, the soccer game, the new intern, Saturday's philosophy, the newborn, the tight deadline, the old car, the record traffic, poetry, bitterness, the concrete sidewalk, the sea, and the applauded sunset, Monday will be the same as all the other ones.

Some pretend that Sunday isn't there. They go to the park, the mall, their mom's house, grandmother's house, the beach house, the countryside…Children fall off their bikes, filmmakers film, writers write, and firefighters put out fires. Some fall desperately in love for 90 minutes; some fall in love for the whole day, and some fall out of love, argue, and leave the house.

In restaurants, the usual group of friends; in the movie theater lines, couples who aren't as enchanted as they were last week; teenagers who ditched their parents at the door; loners with a book; the lone iPad user three people ahead; the couple who met yesterday and continued their outing; the parents who went to see the movie for the third time; and the family that went there this month because they had a little extra money to spare. The food courts were packed, the bars full, the theaters crowded, and the museums bustling.

The afternoon descends with a longing for those far away, the nostalgia of photos, the new book, and the old couch. The TV is on. Everything is the same.

Sunday is always somewhat gray for the soul, even when it's blue for the body.

Untitled

It's no longer news
It's been one year
It passed by quickly
It passed by slowly
It was indispensable

My poetry was not about you

At some point, you wanted to unsettle me
You found me arrogant, pretentious
I looked down on you and you paid for my dinner
I ignored you and you courted me

And you left marks, as you wished
I never forgot the abusive relationship we had
The scars no longer show on the surface
but I was never the same again
I lowered my head
endured other abuses in silence
needed to relearn how to defend myself
I let my guard down for you, and it was difficult to rise again
But it was good to learn that if I need a guard, maybe the relationship isn't good for me (just a tip, girls!)
It was good to include in this book

But nothing of mine was about you
It was about what I let dive into my sacred
You were a fishing instrument against the current
The bridge I built
with the debris I sought
at the bottom of the river
All of me, by me, for myself

I forgave myself in tears
in the profit of Heineken
in the darkness of the room
in the reinforced patterns in solitary recovery

It's not you, it's me
    Everything in São Paulo was new
    It was the first time I felt the world
    I wrote to myself about the new way of observing the familiar

    free

    from my Rio
    my place
    my home
    my mother
    my biased view

    Possessive pronouns born outside of me
    in the withdrawal symptoms
    of undiagnosed crises.

São Paulo has something that, to this day, I can't explain

In 2018, I came to New York. I made friends here, am building my roots, and am happy, but I remember São Paulo from time to time. The other day, I saw a photo of a makeup artist friend on Facebook during a job: the plus-size model with orange hair was wearing a half-sleeve dress in a backyard with a blue wall, and I thought:

– This is so São Paulo! And I came here to write to understand exactly what that "it" is.

São Paulo has a freedom I don't remember seeing in Rio, which often lacks in New York, not because someone is "judging you," it's more of a physical thing. When the weather is extreme, the external limits you.

In São Paulo, I was free, and maybe this freedom is only in my memory, or maybe it's the city's dynamic "hustle." At the end of my four and a half years living there, I was already tired and thought there was nothing left to explore in the city. Deluded is the right word, right?

São Paulo is a gateway to Brazil (all of it). I think that's the magic. At the same time, it's somewhat unfinished. Even in its mansions identified by bank totems and fast-food chains, it has an open ending, like this chronicle poem.

City Echoes: lessons from New York

Perhaps it's strange that the part of the book in which I write about the city where I lived the least is the longest. But it was here, in New York, one of the largest metropolises on the planet, where everything - absolutely everything - happens. Right here, I discovered myself as a writer. I accepted myself, hit my chest, and said that this is a profession. I circumvented all the ruminations of my mind to finish this book and the other millions of ideas I have already organized in special folders. And, if it was here that I built my confidence, it is only fair that I write more about the place where I live, reinvent and accept myself as I am, shedding the beliefs I left on the plane that brought me from Brazil for the first time, praying not to infect the passengers around me with doubt.

I never thank New York enough for making me who I am now. I think I left some of the fantasy in the punches I took in São Paulo, which I'm grateful for. Enough now. It's time to shine!

The Day I Entered Limbo

Once, I wrote in a blog post about the four phases of adaptation abroad: survival, disillusionment, limbo, and screw the limbo. We don't always have a clear start and end to each phase because life is not linear – you may interlace them at times. I think that's what happens most often.

After writing about those phases, I remembered exactly when I entered limbo. I had just passed through my survival phase, and my wife was traveling a lot for work. I was completely alone in this concrete jungle, and a friend from London invited me to spend a few days there with her. Since I already knew my whereabouts there, I thought it would be a good idea to go.

("We were so free before the pandemic, right?")

Here in the United States, there is a thing called a credit score: the higher it is, the more credit you have, the more loans you can take out, and the more postpaid cellular service plans you can have. You build a good credit score by paying your bills on time.

At the time, I had been in the United States for two months and only had a prepaid cell phone. Here, having "good credit" is something you earn, and it greatly influences the country's culture, but that's a conversation for another day.

I went to London without my phone's internet connection and wrote down the directions to my friend's house on a piece of paper. Guess what? I forgot and didn't pay the phone bill.

I returned to New York without a phone. You don't take an Uber when living abroad for two months. I took the subway back home, got off at the closest station possible, and got lost. It was on that day that I entered limbo; it was on that day that I felt insecure even to go to the corner. It was after that day that everything took on proportions much larger than I could handle because not knowing how to get back home is similar to drowning, and no one goes back into the sea after they've drowned.

The first time I had a first date

Throughout my love life, I hadn't experienced many first dates. Actually, I only had one with an ex-girlfriend I met on a dating app.

I never saw myself in front of the mirror looking for the perfect outfit or mentally pondering the topics I could or couldn't talk about with a particular person when meeting them.

I met many of my girlfriends through mutual friends and married a friend, and that was how it went throughout my single life. Perhaps that's why I never entirely understood the fear of the first date, which I saw so much (and still see) in TV shows and movies.

Today, I had a terrible first date. It was as scary as it was funny. It wasn't a romantic date, but a professional one. A friend from Brazil, being kind and helpful—as we tend to be in Brazil—introduced me to an acquaintance who lives here and works in the same field as me. Let's call her Joana.

Joana has been living in New York for five years. She was very helpful in our first contact via Instagram and scheduled a meeting two days later. The chosen place was a restaurant where (I imagined) we could have lunch and discuss the job market in the northern hemisphere.

I arrived, and she asked me about my career trajectory. Naturally, I told her I've been working in the same field for over ten years, explained the projects I used to do in Brazil, and mentioned the international jobs that gave me a foundation to understand processes in other places. At that moment, Joana started treating me like an enemy, and I began to question the purpose of that meeting.

I was sitting at the table with someone who claimed to know all types of visas, including those for "Canada and Australia," but didn't know mine, which made it inferior or subject to questioning, to say the least. The peak of the conversation was when I showed - on the U.S. government website - the type of work authorization I have, and she insisted that I might not necessarily be able to work in our field.

Her resistance made me think about what people expect from us. Did I say something that offended her? Was I too straightforward, and she was expecting someone more timid? When we move to another country, do we need to create a less aggressive personality to be better perceived by our fellow countrymen? Did this specific Brazilian get bitten by the bug of pretentiousness and forget that she once had no experience or needed someone to pull her along?

I am far from being emphatic, but I also don't usually embellish this kind of situation. I was expecting a meeting where I would come out understanding a bit about the entertainment industry in New York, and I heard vague answers like "It depends on the type of work," "I'm sure, but some concepts are static, even in such a broad industry like ours."

The turning point came when I explained the process in Brazil in detail, and she gave me information I could have found on Google. I felt like I was talking to a wall.

Her expression changed when I mentioned the resistance I feel in the United States, especially with foreigners. Something inside Joana screamed and jumped out, and she clearly wasn't willing to share her struggles with me. I sympathized. I thanked her for her time and left, replaying our conversation in my mind, unsuccessfully searching for the reasons for such a traumatizing encounter for both of us.

I went to lunch seeking answers and came back with more doubts, especially about the lack of empathy that makes us disregard the humanity in front of us. And hungry, I left the restaurant hungry. The 40 minutes we spent face-to-face were strange. Perhaps I will still have many lousy work encounters, but this one will forever be my first.

Monsters in New York

Here I am again, at the bottom of the pit.

And this time, I can't even blame my romantic relationship. That's the one area of my life that's going right.

I was so directly confronted with the monsters that I had tried to hide for so long that I fell at once and, without realizing it, was leaving my nails in the walls.

What is this addictive personality that takes over me?

Every day
my body tries to expel my demons
that resist bravely.

I only find peace when I write, soothing what eats away at me.

It was time not to feel anymore

I celebrated my 19th birthday in London, as far away from home as possible, as I still do today. I left my teenage years with deep scars and decided that it no longer made sense to confront the world in the same way, facing monsters. I decided that it was time not to feel anymore.

– From now on, I will be happy and never feel any pain again. I will take care of myself, not depend on anyone, not get attached, and not suffer again.

I should have considered - again or for the first time? Caring for myself was only possible by making those old scars burn.

Not feeling pain is only possible when we treat the pain and the treatment hurts. I trained my monsters by closing deep scars while gaining superficial ones. What I didn't understand at 19 was that to tame sadness, I needed to face it head-on, and doing so meant freeing myself from the cages I had built to protect myself. Being free hurts.

Not belonging messes with our heads

Belonging is a harsh feeling, right?

We have been seeking it forever and always (I think; I seek it).

The other day, I read—I think it was in Maria Ribeiro's book (a Brazilian author), which is by my bedside—that we are a bit of each person around us. Writing it like this even seems like a topic from somewhat legendary studies, you know? It's the kind that everyone has heard of, but no one knows exactly the source (be careful with the "legends" when voting).

I looked it up on Kindle and found the excerpt from the book. It is indeed Maria Ribeiro's idea that I am trying to reformulate - may she help me:

– I dated Rafael and became a bit like Che Guevara, then I dated Paulo and became totally MR8 (a revolutionary leftist guerilla)

Asking Maria's permission - Maria, I, too, was a chameleon in my relationships, and more than in relationships, I was a chameleon (sorry for repeating words) in the cities I lived in.

In Rio, I had friends who swore they had never seen me wearing pants. In São Paulo, I wore them all the time. Now, in New York, I am still trying to adapt.

I don't like wearing pants, but the text isn't about that. It's about this feeling of not belonging that messes with our heads.

I once heard that a shy person thinks everyone is looking at them. I wish I had heard – or accepted? – that 15 years earlier.

I live as if someone – or everyone – is expecting the hot dog I eat in front of Central Park to fall on my white shirt and mess me up or that I switch seats on the subway, slightly bothered by something imperceptible, to be accused of indecisive snobbery.

It's confusing, isn't it? I'll try again: I feel inadequate and think all eyes are on me, like the actor in the spotlight on stage who forgets the lines or when I arrive at school without clothes and become a laughingstock.

– Have you ever felt like this, Maria?

I felt like experiencing what I didn't experience

A person in New York doesn't live solely on mirror selfies. Thankfully! Yesterday, I visited one of the city's most beautiful places I've ever been to.

I felt like going back to school
to have made better efforts
to drool over the teachers' wisdom in class
to be 19 again (oh, goodness, no! But I wish)
to have had different experiences, without ceasing to be myself

Columbia's campus has an intimate touch to its magnificence. The echo of the traditional library hall is welcoming. Names and monuments of great minds (unknown to me) invite us for an immediate trip into the utopia of infinite knowledge.

I was enchanted not only by the architecture but also by the words of Clima Change Brazilian Activists Celia Xakriabá, Petra Costa, Eve Ensler, Gabriel, and Glenn, who restored my ability to see colors in a horizon of dark clouds.

In the final act, we sang with Caetano, a green sigh amidst the flames—all of this for free (0800). My only goal was to feed my soul with hope. #keepgoing

Essays from a Sunday night

Sunday, September 1, 2019. Over four years ago, my wife and I boarded a plane headed to the biggest change in our lives, or New York, for those who prefer a pragmatic life. This is not a pragmatic text, and it has sprinkles of ice cream and love all over it, which are almost the same for me.

Back to Sunday: we arrived at the restaurant I chose for lunch around 4:00 pm. We were greeted by an unpleasant sign that said:

– Closed to the public.

The sign next to it asked wedding guests to use another entrance to the establishment. The bride and groom chose the place well, and we were the ones who erred in not calling ahead. In New York, there's still this culture of making reservations at places, no matter how common the restaurant is. It reminded me a bit of the 1990s in Brazil, even though it was 2019.

People in New York are already used to bumping into others in the street and gesturing alone with earphones. They talk a lot on the phone—it reminds me of a production room—much more than we, Latines, who prefer to text. Sending a long voice message on WhatsApp is considered rude in Brazil, and the sender even apologizes afterward, of course, through text.

My wife Marcela and I were showing signs of a bad mood. The signs in front of the restaurant did not help at all, but they were not to blame for our bodies' reactions. The actual cause was the significant amount of time since our last meal: bread with butter at 10:00 in the morning.

We called an Uber to go from home to the restaurant, and on the way, I thought about the quickest appetizer I would order to calm my growling stomach. I needed to change, to adapt; that sign saying "Closed for a private event" could have ruined my day. We walked hangry to "kill what was killing us," as my mother would say.

We stopped at Time Out, a kind of upscale food court, a concept quite different from what I had in mind when I spent 20 minutes researching Google Maps for a pleasant option for the palate, eyes, and wallet.

We waited for about ten minutes to be served at the counter of a sandwich shop, where eight employees were assembling dishes, but no one was at the cash register. We placed our order and searched for a table outside the food court. Dining with loud music is no longer our thing.

In our hands, a kind of pager that vibrates when your order is ready or our passport to a better late afternoon.

Also, I wouldn't say I like good taste

Finally, I understood why Adriana Calcanhoto (a famous Brazilian singer) doesn't like good taste. Last Friday, I was irritated for a millisecond by the noise and joy that a colorful Latino family brought to the subway I was in.

The path was silent before, but as soon as the two couples of cousins, friends, or siblings with children, teenagers, and young adults entered, the view became noisy and colorful.

How strange is it to like silence, absence, and limits when everyday life is confusing, noisy, and full of colors? I wanted silence right there in New York, where everything vibrates; it's hectic and rushed.

New York is a city of fearless people

New York is a city of fearless people, or at least people who pretend well that the exorbitant rent doesn't weigh on every decision made during the workday.

New York is a city with big companies and one of the least entrepreneurial in the United States, yet everything happens here. Silicon Valley, excuse me, but if it weren't for New York, you wouldn't even have made it there.

I'm not saying this as a nostalgic person from my hometown's nobility. New York doesn't live on old money; it thrives on new ambitions.

New York is a city of fearless conquerors. How could I, who was afraid of everything, feel like I belonged? On the steps of the MET, having conquered nothing, I see people from all over the world with fewer belongings than me, unafraid to be who they are, unlike me.

– Redheaded lady, do you also feel inadequate?

Today, the One World no longer scares me

It's easy to dream of New York to live or visit one day, but living with the One World staring at you from the living room window shouting:

– I doubt you can do it. - is a more thorny reality.

When I arrived, I walked the streets feeling frightened, not because it was dangerous but because it was an imposing city.

If you've never felt small in life, you've never stepped foot in New York. Here, the skyscrapers are intimidating. You, minuscule on the sidewalk, trying to keep up with the hurried pace, look up without hopes of reaching the end of the vastness of concrete in a city where even the sky is grand.

But the sky is a topic for another day because today, the One World Trade Center - with a first and last name - doesn't scare me anymore.

New York has everything, except an explanation

New York is a unique city with an absurd diversity of people, cultures, places, lifestyles, and everything that exists in the world.

There's a saying I always use when I miss something very Brazilian:

– If it doesn't exist in New York, it doesn't exist anywhere else in the world!

Said and done. Everything is found here. Before living in New York, I wanted to know why this city was chosen to be the setting for so many movies.

What would be so unique in the corners of the streets here? And that's it. There is no correct answer. The air here has something that pushes, challenges, and makes it seem like everything is possible, even if it's not true.

I walk around New York, and I see movie scenes, street basketball games, piers overlooking the world's most recognized skyline, green parks in the summer, tall buildings, bustling streets, and empty houses with their hastily closed curtains, where you can always see a glimpse of the makeshift basement apartment in a home without a garden.

Every end of summer

Every end of summer, I feel like I didn't enjoy it properly, as if I could hold on to the time and space of summer, the summer photo, the summer lake, and the cold seawater in the hot summer. I want to hold on to summer and stay forever. No winter - no weather discomfort will be tolerated - as if I could slip by and go back to the previous summer, stuck at the same age, the same innocence, the same lack of commitment, without having a birthday.

Not knowing what you want for the future is never cute. You can't even write, and they already ask you what you want to be when you grow up. That's how it was for me. Was it the fear of not having an answer that paralyzed me at five feet and one (and a half) inches? I never grew much, and I never had an answer.

Not knowing what you want in life at 33 is horrible; it's judgmental and a guilt trip. As if life ended last summer, when we were still young and drank beer on the hot sand of Ipanema in the early morning, as if not knowing K-Pop was a death sentence; as if, after 25, everything had to be resolved and static. No winter, summer, and nothing in between.

In most parts of Brazil, mainly in Rio, there is nothing in between, not even winter, to be honest. It's 365 days of summer I and II, with a few cool hours. Did I bring from Rio my desire to only live in the summer? It seems too obvious. They say clichés exist because they work. A supermarket cashier in Sheila Heti's novel says that women always make their lives harder. Maybe the obvious works.

Headlines emphasize that "someone entered college in their 50s," as if there was an age to do anything. Leave your parents' house, become an adult, work with what you want, and study when you can. There is no deadline to do anything. Becoming an adult with the support of parents is great, but not everyone has that privilege for reasons I can't begin to list.

Maybe the proofreader of the Portuguese edition of this book - who said I should've written more - would find it a good idea for me to list all the reasons someone has to leave their parents' house early, but I don't think so. I would feel kind of childish in the summer of 1998, procrastinating on the composition task.

If you have the privilege of living with your parents, enjoy it. Your youth can be freer. Following the paths already traced increases self-confidence. There's nothing wrong with wanting what everyone else wants, but nothing is guaranteed except for the guaranteed return trip to Brazil every summer.

Disconnection NY-BR

Lying in a meditation position during the online yoga class I take on Mondays and Thursdays since March 2020, I felt the cold that used to blow on my back during the late-night shoots I did in São Paulo. My teacher, on the other side of the screen, and the Equator, was bothered by the heat.

Autumn in New York is precisely like winter in São Paulo. Some days are warmer, others colder, and the certainty of needing layers of clothing and an umbrella (never forget your umbrella) to spend the day outside.

When the foliage here starts to lose color and the fashion there begins to lose fabric, I feel far from Brazil. Even with the one-hour time difference, I feel disconnected, preparing for a dark winter like the last five. At the same time, I see my Brazilian friends on social media preparing for unbearably sunny days.

In 2020, everything was worse with the pandemic, of course. As winter approached, New York was getting ready to close some neighborhoods again, and the possibility of spending the end of the year in Brazil seemed increasingly distant.

May "Saint Cher" help me face the upcoming winter —we say face because it is indeed a struggle to withstand the negative temperatures of the short days.

Tourist at home

Before feeling at home, I walked the streets trying to record every detail, like tourists or sad people, clinging with all their might to the only breath of relief, which barely illuminates the depths of their lives.

When - finally - will I feel like a local?

I remember getting to know Rio (no true *carioca* calls their city Rio de Janeiro), where I was born and from where I only left just before my 25th birthday, heading to São Paulo without claiming the return ticket until today.

In Rio, I knew the ways between my house and my father's, my grandmother's, the school, the shopping mall where I took English classes, my mom's friend's house in Copacabana, my friends who lived in Tijuca and Recreio, my college and my friends' colleges as well, and all my jobs, even if they were in not-so-familiar neighborhoods.

In childhood, I knew every street corner from my school, Barra to Alto da Boa Vista, and was welcome almost everywhere in that city. I also had in my mind the coordinates of all traffic lights and the time it took to get from one destination to another.

The experience of the city by car and on foot is different. Here, in New York, I don't drive much, although I already know there is an express line - like the highway in São Paulo - that cuts through the city with water views - along the East Side - and that the bridge near my house leads directly to Chinatown.

Strolling around Central Park last weekend, I saw some small tunnels for cars and wondered: can cars drive through the park from east to west? Is it underground? Where do they come from, and where do they go?

Besides that, I still don't know exactly where I am welcome in New York, but that's a topic for another day. Today, I want to say that I have been living here for two years and still don't feel like a local. Will that day ever come?

P.S: Yes, cars can drive through the park, and no, they don't go underground. I found out on Google Maps.

The Shrinking of the Immigrant

It seems contradictory to discuss the shrinking number of immigrants when they arrive in a new country to expand.

In the almost four years I have lived in New York, I have met many people, especially women and LGBTQ+, who came from Brazil to the United States in search of a better quality of life.

Everyone has their own definition of "quality of life," but improving is somewhat synonymous with change. If it weren't, everything would stay the same, and it would be fine, but that's not the case.

I don't know if we think properly—I didn't think, and honestly, I don't think it's possible to measure the size of the change before actually boarding the plane and moving.

The truth is that new opportunities are sought with the filter of thinking you know the future. It is very difficult to predict the challenges of a new city, even if you know it well.

When I moved to São Paulo, I felt overwhelmed. My arrogance made me believe that SP was the backyard of my house. But the city is gigantic and wouldn't fit in my wildest dreams. I think it's like that everywhere.

Nobody knows all the secrets of themselves, imagine the places they live.

I know people who worked online before coming here and had difficulty adjusting professionally, people who were transferred, children of Americans, and "locals" who didn't find the change a walk in the park.

The truth is that "expanding," for most immigrants, means having more cultural, financial, professional, and intellectual access. But this expansion comes at a high—almost oppressive—cost.

Leaving your homeland means confronting every day with unimaginable challenges, incomprehensible vocabularies, customs you don't have, and foods you can't find. Moving to another country means immersing yourself in a culture that you only knew through the lenses of cinema or memory.

Instead of expanding the minds of those who arrive, all these facets seem to retract. It is painful to relearn how to live after having an established life.

(Re)learning shakes your core, self-confidence, self-esteem, and the security you have in yourself. Assimilating a new culture makes you question everything that makes you, you.

The shrinking of the immigrant comes from almost everyone. I have met people with the most varied social certifications, shrinking. I saw them fade. I saw myself fade.

I witnessed people questioning established values and shaking firmly rooted foundations. The word "expansion" is accompanied by many details, which we can also call tolls. And the price you pay to have a better life is the time it takes to clear away everything in you that is not yours, to recover, and ultimately, to expand.

I saw all of this in less than four years. During this time, I became an external and internal immigrant. If you want to make this journey to escape something, be my guest. I cannot say I didn't do the same.

But know that moving countries is like diving into an image.

Poem for New York

The poem was written especially for the Love Notes to NYC project at the invitation of its creators and curators. The original poem was written in English, following the project's proposal.

> NY, I was not ready for you
> I think no one is really ready for you
> You are an intimidating concrete jungle
> And before you come into our lives, we are helpless babies
>
> Nobody comes to NYC; it is NYC that comes into our lives
> It invades our bodies with dazzling lights of opportunities
>
> We

Larissa Rinaldi

paralyze

Wait

Breath

I should have never left home
We are grown-ups
on the outside only
Inside, we feel tiny,
looking up with a pacifier in the mouth
dizzy with the rush of your streets

But you got tiny, NYC
Your buildings are still here, looking upon us
But you, you got small
Waiting for an answer
You, who knows it all, were waiting for us
to take care of your empty streets

How did it feel, NYC, to feel small for a short time?

New Yorkers took care of you,
We've struggled to make you proud again
And so you are

Proud of us
Proud of your people from all over the world
Proud that you taught us well

You are standing up to be a giant again

Above all, you are proud because you always knew
That your greatness is on us
And we didn't let you down
That's how we say: we heart you, NYC

Here is the translated version in Portuguese:
*Nova York, eu não estava pronta para você
Eu acho que ninguém realmente está pronto para você
Você é uma selva de pedras intimidadora
e antes você entrar nas nossas vidas
somos apenas bebês desamparados*

*Ninguém vem para Nova York,
é Nova York que entra em nossas vidas
Invade nossos corpos com oportunidades que confundem e ofuscam*

A gente

paralisa

espera

respira

*Aqui, nós parecemos gente grande
mas nos sentimos minúsculos
admirando o horizonte entre os topos dos prédios, chupando os dedos*

Larissa Rinaldi

completamente tontos com a pressa de suas ruas

Mas você, Nova York, você ficou pequena
Seus prédios ainda nos olham de cima
imponentes
Mas você ficou pequena
esperando por uma resposta

Você, que tudo sabe, esperou por nós
para tomar conta de suas ruas vazias

Como foi se sentir pequena?

Nova Yorkinos tomaram conta de você
A gente lutou para que você se orgulhasse de novo
e conseguimos

Você está orgulhosa de nós
Orgulhosa do seu povo cosmopolita

The End and the Beginning

Post-pandemic texts

My solitude, your solitude

For almost a year, we were isolated, didn't hug our friends, sent kisses from afar, sent hearts through the screen; ten months…

I confess that in March of 2020, when all this began, I thought that finally my friends in Brazil would have time to talk to me and that this would be a good thing, as I missed the deep conversations on sweaty Summer nights with no end time.

Ten months later, I can affirm with all certainty that there was nothing good about this pandemic. I developed my profession, it's true, but nothing compensates, not even close, for the deaths and the lack of freedom we experienced.

In March, I had the illusion that everything would end quickly. In the middle of the year, I would see my friends on a beach, and in December, I would take a flight to Brazil, as I had planned.

The vaccine is here today, but I have less desire to make plans. Why, right? If 2020 taught us anything, it's that planning doesn't always work out, but it's sorely missed.

All That Was Known

While we waited for hospitals to become less crowded and scientists to bring the vaccine or effective treatment, we stayed at home, without contact with other people, without crowding, without working, without going to the cinema, museum, bar, parties, or traveling. We breathed on machines in a figurative sense, while millions needed ventilators in a literal sense.

In the last nine months, we did what we could to avoid going crazy dealing with the so-called new normal. Some messed up, and we got angry—and still do. I do.

In 2020, I had a lot, but a lot of time to look within and understand who I am amidst so many internal and external revolutions. I learned to define priorities, set objectives—almost always clear ones—and, most importantly, trust myself. I finally understood that it's okay to make mistakes and keep trying, change strategy, ask for help, share less--than-good moments, appear vulnerable, write about it, and publish.

The journey I made within myself was crazier and more transformative than any other I could have done. That's all I knew about 2020.

Will New York ever be the same again?

The last year was terrible for the whole world. We saw New York become the pandemic's epicenter in March 2020, and since then, we had to get used to a city that was anything but the New York we knew.

The summer, Halloween, Christmas, and even Times Square were empty. On the streets, there was the even more deafening sound of sirens and muffled cries for justice. It wasn't just the city that emptied and got lost. We, the residents, also lost hope every time we saw our favorite restaurants close.

For months, we lived in a black hole, but the vaccine arrived, and things began to change. It didn't happen overnight, but they started to change. With the arrival of spring, after a winter that - felt like - took a year to end, and with the progress of vaccination, gradually, I saw more life on the streets.

Gradually, New York returned to being itself.

Nova York, aos poucos, volta a ser a mesma.

Everything except dead

As Madonna sings, "I don't like cities, but I like New York," and I understand her point of view. I am a child of a big city, I never appreciated the simple life, nature, or the lemon picked from a tree on my grandfather's countryside estate.

I confess that the more urban I became, the more I valued moments in contact with nature. It's easy to be a city girl when you live in Rio de Janeiro, right? In São Paulo, the story was different, and in New York, my relationship with nature changed even more.

Nature is different, the birds are different, the vegetation is different, and the marked seasons of the year change the city's landscape every three months. Having a relationship with nature here means having a different relationship with your wardrobe.

I wrote all this to say that much has been said about the death of New York during the exodus that occurred during the pandemic, but if we once saw an empty, sad, lifeless city, that city stayed in the past along with the fallen leaves of the last autumn and the dead trees of winter. Today, what you see is the city that made Madonna become Madonna.

Frantic, bustling, dynamic, noisy, rushed, with its executives in suits timidly returning to the streets, its exorbitant rents, the strong smell of different cultures that run through the entire city, the accents, and the uncontrollable desire to shine in the midst of all of this.

New York didn't die because it remains the world's greatest cradle of diversity, and diversity brings creativity and innovation. Here we are, New Yorkers, without putting limits on dreams, ready to conquer the world again.

The Spring Pranks

It's hot. Not really! Every transitional season - spring and fall - is the same. It's 20 degrees one day and zero two days later. Putting away the thermal clothes to hibernate in the bed trunk is an act of:

– Enough! Winter is not coming in here anymore.

But it does come in, and we feel cold. New Yorkers don't care when it's 32° because they've experienced -4° degrees and the sun setting at 4 pm.

The streets were crowded again; vaccination was progressing; flowers were blooming; jazz musicians were back in the parks; restaurants, at 50% capacity, were full with a mask-wearing line at the door. Life slowly returned to the city that spent so much time in the silence of solitude.

– Come, Summer, New York is ready to embrace you!

36 million people

The other day, I read that the city of New York invested 30 million dollars in campaigns to bring back tourists and expected to receive over 36 million people in 2021. Thirty-six million people! I swear I had no idea about the size of tourism in New York until I came across those numbers. In 2019, the city welcomed over 66 million tourists. Can you believe it? Neither could I.

Just in 2021, over 36 million dreams, hopes, anxieties, fears, joys, parents, mothers, children, spouses, siblings, cousins, boyfriends, girlfriends, singles, young and old were expected in the city of New York.

Just in 2021, more people passed through these streets than the entire population of some countries.

Many said that New York died during the pandemic, but if it did, it rose from the ashes like a phoenix, powerful, ready to impress the 36 million hearts that pause for a millisecond with the grandeur of our buildings.

Welcome tourists, and may we be able to welcome them safely.

P.S.: Since we work with data here, here is the link to the ABC article with the numbers mentioned in this chronicle. Available at: https://abc7ny.com/covid-vaccine-nyc-tourism- -new-york-city-reawakens- -mayor-de-blasio/10534680

https://abc7ny.com/covid-vaccine-nyc-tourism-new-york-city-reawakens-mayor-de-blasio/10534680

It's allowed, it's all allowed!

Ready or not, the masks were coming off. Following the recommendations of the Centers for Disease Control and Prevention, the governor of New York lifted the mask mandate for vaccinated individuals, with a few exceptions. Finally, after such a difficult year—and indoors—New Yorkers will be able to enjoy the Summer of 2021 with a sense of normalcy.

I'm curious if there are few vaccinated people or if, like me, many still feel strange without a face covering, but what I see on the streets is a certain resistance to removing the mask. Last Sunday, my wife and I walked for about ten minutes without masks. It was a bit strange to control my facial expressions for the first time in 14 months, but it was comfortable to walk freely, feeling the spring air entering through my eager and uncovered nostrils.

In March 2020, we trusted science and stayed home, even without knowing much about the virus. It's time to trust again, even with such a new vaccine.

Whether or not it is strange to walk without a mask after the vaccine, it is allowed.

Sendo estranho ou não andar sem máscara, está liberado!

Source: The New York Times. Available at: www.nytimes.com/2021/05/17/world/new-york-masks-cdc-vaccine.html

Your Future Image

In this life of changing countries and being an immigrant, I am once again seeking to establish myself professionally. Last week, I received the news I had been waiting for for a year: I am eligible to work in the United States again.

This was the end of a journey that began amid a global pandemic, which I shared with many immigrants. During the year my application was being evaluated, I was evaluating myself, preparing for the moment when I could be in charge of my own destiny again.

But I forgot that the end and the beginning are always connected.

During the pandemic years, I looked inward, studied, wrote a book, created a portfolio, produced two podcasts, did everything in my power to develop new professional skills, and, of course, not go crazy.

It wasn't easy, but I did it. All these moves strengthened me to start another journey, during which I learned how to navigate a new world.

I look at my friends around me and see the choices I didn't make. I try to quiet my anxiety by rationalizing that two days are not enough to see results, nor are two months.

I calm my petulant child and seek to develop a beginner's mentality:

How can I move forward from now on? I still don't have the answer.

I persist in non-obvious paths, knocking on closed doors, but that's life, right? You make space for your allegory to pass. You are the driver and highlight of your float. You project onto yourself your image of the future.

.

How should we be?

When does a person realize they have become an adult? Maybe I'm a little old for this question, but it's strange, right? Seeing yourself as a 100% functional human being. Not needing your parents for anything.

I am entirely responsible for everything I do in the world (and you too). I don't want to create panic, but I'm panicking a bit.

I have a friend who has been wearing the same Converse Chuck Taylor All-Star sneakers since we were 15 years old. When her sneakers get old, she buys a new one exactly like the last one. Even the Converse Chuck Taylor All-Star has gone through a brand makeover and now uses 500 names for the sneakers (the most worn-out in the world) that, in my time, were only called All-Star. My friend has not changed her taste in sneakers for 17 years. I think that's the most adult thing someone can do.

I do not have a favorite sneaker to this day. I had phases, but I changed so much that I don't even remember why I liked All-Star in my teens. I know that teenagers are discovering the world, testing everything as quickly as possible to decide what will remain for the rest of their lives.

Sometimes, I feel like a teenager discovering a whole new world when I need to buy sneakers, clothes, or anything else, actually. I don't have a favorite anything. I may have a favorite color, but that's it. I don't have favorite sneakers, bags, magazines, restaurants, or places. My favorite thing to do is change my mind.

My constant moving around didn't help me establish favorites. I had to adapt, and every city has its own demands. How can someone have a favorite shoe while living in New York (with temperatures ranging from 32° to 80° in 12 months)? Some wear Timberland boots all year round, but I'm more of a *Sex And The City* type. But it's impossible to be SATC while living in New York. In real life, we walk miles a day and use the subway.

This year, I read a book by Sheila Heti titled How Should a Person Be? Of course, the author doesn't have an answer to that question. Nobody does. We follow standards and rules, hoping to belong to society, our communities, and our homes.

Some people don't follow any rules, some choose which rules to follow, and some use the rules but don't follow them. Rules have their own variables. Exhausting, right?

I think rules are a practical way to figure out how someone should be. Learn the rules. If you don't know what to do with your life, go to school, go to college, get a job. Those are fair rules. You can try to understand who you are while doing all of that. Maybe you'll find good teachers and mentors along the way, and voila! You've discovered who you are. Maybe you're not so lucky, and that's okay too.

It's normal to get lost, never have a favorite sneaker model, or feel angry or lonely. Sometimes, we just need to accept that change is the only constant in life.

The world is a place full of guidelines, but how you should be in this world is a rule that only you can define. It's an especially difficult task because we spend our lives trying to meet the expectations of our parents, friends, society, industries, and our own, of course.

To not end this text without leaving anything behind, I'll share a secret, which might be the only thing I know about how a person should be: we are not just one thing. We are a combination of DNA, history, knowledge, and desires; we are shaped daily by advertising, news, work, and friends. We are the sum of everything we experience daily and the subtraction of everything we are not.

The Human Factor

So, I learned to love myself, gave myself permission to write, sat my butt down and wrote, right?

Wrong. Between my passion and my goals, there's a factor over which I have little control: myself.

I developed my writing and self-confidence to showcase my work, but there isn't a definitive guide on being a successful writer (many try, but none are definitive). It's hard to find mentors (I, for instance, read), and there's no ready-made path. Each writer paves their own way.

There's no rule. There are multiple possibilities and combinations of factors.

– And, damn! I love rules. I love knowing what to expect. I love certainty. I like to see and touch the things I do.

That's why I worked in film production for so long. I could see the movies I made and discuss them with my colleagues. I could touch the value attributed to my work by buying things. However, being a writer is a bit different.

Being a writer means dealing with rejection thousands of times before publishing a work, and the first one to reject my work is myself. First, I find my ideas a bit silly. Then, I work a lot, but I never find the project good enough to be published.

After some months (or years) of working, I convinced myself that it was time to try to sell, but I never achieved the miracle of publication. It's hard for everyone, and each person deals with frustration differently.

I'm reaching the point I want to make: the 100% immeasurable factor that guides all of us, the human factor. Before achieving anything, we must deal with our insecurities, imposter syndrome, past experiences, culture, and other unique social and psychological aspects.

I know how things work. I know when I'm procrastinating when I'm not willing to try to sell my work because I'm feeling a bit depressed, when I'm less productive at certain times of my menstrual cycle. I know the Pomodoro technique. I go to therapy, exercise, meditate, read, and write daily.

Occasionally, I try to sell my work, and rejection doesn't kill me, but knowing how things work doesn't change how I, a unique and full of confusing feelings person, feel after each "no."

Knowing the rules doesn't change the fact that I feel everything anew when I start over because not getting the expected result makes me question the process every time. The point is that there's nothing else I want to do in my life, so I convince myself that I'm learning and carry on.

I used to feel sorry for myself and thought:

– Oh, Universe, why wasn't I born "normal"?

Why write? What a curse for someone so insignificant, the curse of a thousand ideas. Oh, goddess, why?"

I don't fall into that cycle anymore. No, because I understand what leads me to conduct my writing my way despite all the advice.

Knowing the "formula for success" doesn't transform the person who has to sit down and execute the formula. I understood it's not about the formula but how I can deal with my protective instincts to put my work into the world.

Anxiety, fear, and shame won't go away, but they won't paralyze me anymore because now I understand how these feelings work in my creative process. It's different.

It's confusing. It's the human mind that is 100% emotion, but you got it, right?

Breaking patterns

I spent my whole life rebelling. Nobody likes rebels, and to be honest, sometimes I even get tired of rebellion, but today, I understand that my rebellion has an almost vital function. It was a mix of personality and survival instinct. I grew up in a dysfunctional home with little to no space to be a child.

Perhaps that's why, for a good part of my childhood, I focused on learning the patterns that caused pain in my home as if I were sculpting a library of feelings, reactions, shapes, and thoughts and building a file to look at from the outside—a mental space to consult before being the cause of my own suffering.

But living happens in the events that are not on the agenda. Living happens when we get distracted trying to relieve the weight of obligations. I thought that rejecting patterns was enough to be different. In innocence, I thought denying was enough not to absorb, but being different is not repudiating.

My parents were the most important people in my life until recently. Disagreeing with what they did doesn't mean not absorbing patterns. Behavior patterns are recorded, like tattoos on the body. Once a behavior (physical or mental) is established, it's very difficult to readjust. Maybe it's some kind of protection instinct. Maybe it's hard to change because - contrary to what innocence does believe – growing involves pain. Growing consists of ripping off the soaked band-aid and stitching up the wounds alone on a deserted island.

Achieving a different life in the future is a choice that needs to be made in the present, a process that is usually painful. Not wanting to repeat the patterns I knew in childhood is not enough. I need to understand how each one installed themselves in my system. I need to observe, talk, analyze, reflect, and change.

What I'm trying to say is that either I commit to fighting today, or I will have a life exactly like the people I don't admire, even if it seems like I've made different choices.

The denial phase

AThe worst thing about adapting abroad is... adapting.

I don't know about you, but I don't like change. I like things static as if they weren't changing all the time. Maybe that's why I don't want to have children. Children change, they grow. They don't stay at the same age. They are constantly moving through phases. Everything is a phase when you're a child. There's the "why" phase, the literacy phase, the phase when teeth grow in, and the phase when teeth fall out.

Maybe wanting to stay in one phase is a bit childish as if you loved the previous phase so much that you wanted to hold onto it, but you can't. You must move forward because you won't be six years old forever. I think that's the phase I'm going through now - I want to be six years old forever. Not literally, but figuratively. To adapt to a new culture, you need to understand your limitations and be patient. Children can't do either of those things.

I arrived in New York at 29 years old. The Saturn Return phase is about settling scores with reality. I didn't like dealing with reality much, not because I'm a writer, but because I think it's the other way around. Astrologers say it's at this moment that you learn about your limitations. I found thousands of them in the last few years. It wasn't hard to find limitations miles away from my comfort zone as I tried to adapt. I spent three years in denial, like a stubborn child.

When I was a child, all I wanted was to be an adult and free. I just didn't know that freedom is another word for responsibility. They don't exist without each other. Being free requires commitments and sacrifices. Freedom takes everything from you, shakes your pillars, and turns your world upside down.

Make no mistake. I have no idea what your life will be like after it turns upside down. Maybe you should learn yoga. People say that life is not linear; it's a spiral.

What the heck does that mean? Am I going back to the Saturn Return? I don't like that idea at all. I wish that moving to the next phase in life was like graduation - that it would come with a diploma to prove you're ready for the next phase.

But life has no guarantees, and moving to different phases doesn't come with a signed document with a recognized signature.

I wanted assurances, but for now, I'll just be very adult and accept that I have no idea what will happen in my next phase. Maybe I'll go back to writing fiction. I miss a good romance. That's probably why people have children. Books about raising children always state the next stage.

Speaking Two Languages

I don't know how non-writers' heads work; I don't know how other writers' heads work; I barely know how my head works—and knowing how our brain works is the goal, right? Making our minds work for our success. Maybe I would be more successful if I had more control over my mind.

But then, we can discuss what success exactly is, and I didn't want to talk about success today. As you can witness, here I am, losing control of my thoughts. Again.

Even though I know practical things about my mind, some facts fascinate me, like being bilingual. Speaking two languages is almost an uncontrollable instinct to need to write in English or Portuguese, depending on my inspiration's desire.

To me, communication is the most beautiful skill in the world. Everyone needs to communicate. Great successes and failures start with our ability to understand and eventually agree with others.

I feel like the luckiest person in the world to be able to read and write. I treat being literate as something extraordinary despite it being basic for most of us who had access to elementary and higher education. Still, I think the ability to read and write is a gift that should be treated with care and respect.

I show respect by reading and writing as much as possible in the two languages I know. I write because I enjoy it, because I find it beautiful to have the ability to put ideas on paper, because I love the sound of the keyboard, and because handwriting is fascinating.

It's not that my handwriting is beautiful. What enchants me is the process of sitting down and writing until my wrist hurts. What fascinates me is letting thoughts freely transform into words, like a spell from a witch's wand.

Sitting down and handwriting is a special moment for me, and I don't care if it seems silly to the rest of the world.

I enjoy speaking another language because being bilingual means having different personalities, and studies confirm what I just wrote. For reference, please read "Becoming Bicultural - Risk, Resilience, and Latino Youth" by Paul R. Smokowski and Martica Bacallao; and "Negotiating Bilingual and Bicultural Identities - Japanese Returnees Betwixt Two Worlds" by Yasuko Kanno.

I feel very privileged to have been able to learn English one day and fully understand situations completely different from those that happen in Brazil. Understanding a whole universe so distant from mine is breathtaking, and no one can take that away from me.

Of course, it's not the best of worlds when I want to communicate and the words I need to make sense of a sentence disappear. Then I feel like an idiot, a fraud. I feel like someone who is wasting the greatest blessing of the world on the translation apps I constantly use.

But translating is part of it. I'm not always in control of my bilingualism. I'm just a player in the fluid relationship that exists between the two parts of my brain.

Writing is giving voice to the soul

The other day, I was talking to a Brazilian/American woman. She told me there was an error in an English sentence I posted and asked me why I decided to post in two languages.

I replied that I did it because I wanted to, because my audience understands English, because half of them also live here, and especially because I don't want to limit myself to one language.

I perfectly understand both languages. I read more in English than Portuguese, and I don't prefer to create in one language or another. My creativity doesn't work like that.

I understand the need for constancy in my work, but she didn't understand that my coherence lies in traveling between both worlds.

Making a mistake in writing is not a problem. I learn every day, even in my native language. It wouldn't be any different with English. What I can't do is stifle my creativity because of a wrong preposition.

The preposition doesn't change, but I can change it and learn to put everything in the right place—just move the prepositions around as I please. Writing is more than perfect grammar construction.

Writing is giving names to feelings and voice to the soul.

I arrived here

I wanted to start the introduction of this part of the book with: "It seems crazy to go back to part 1 after finishing part 2." But then, I saw that's how I started describing the last act and didn't want to repeat it.

I'm having difficulties finishing this book, and I don't deny it. If it were easy to finish, I wouldn't have experienced so many abusive relationships with people and things.

The relationship I have with my texts is a bit abusive, for example. Sitting in front of this screen with a blinking cursor, I feel immense pleasure. I like the sound my finger makes when typing, the silence of the heater on, the aura that protects me from everything happening outside the mind-fingers-screen perimeter, and the agonizing knot in my throat that slightly tightens the jaws relieving tension with (sigh of relief) a scream of pleasure.

So, of course, I don't want to finish this book. Who would willingly end such an intoxicating, intense, vibrant, and, best of all, so uniquely yours relationship? Yours and no one else's. This book, in particular, doesn't want me to finish with it because then I open space for other toxic relationships, and it doesn't want to lose the pedestal and the sovereignty of being the only one capable of making me on the verge of climax at every moment.

Finishing is necessary. Maybe I shouldn't have let my last therapist break up with me. That was a civilized end, the only one I had.

I climaxed. I'm ready for another. Let the next ones come, and may I protect myself from all the fears of being the author of only one book. May the desire to hear the sound of the keyboard remain in me, and may the ability for multiple orgasms endure. Amen!

Every time I left home

Leaving home, the state, and the country. Leaving everything comfortable, everything that fits, and everything familiar. Leaving the place where you know all the distances and measurements. I was going without a ticket back. It's the second time I'm going without a return ticket.

Only those who—before leaving—already feel like strangers to themselves change countries. And I'll go further: only those who question values, seek novelty, and feel a bit rejected in the family environment are somewhat strange and somewhat foreign. Leaving home is not an act of bravery; it's for those seeking something inside that they only find outside. That's why some stay while others are restless.

I've always been the restless type. The other day, an aunt commented that I always said I wanted to live outside Brazil. What was I seeking as a child to want to live abroad? I have a few guesses, but I'll leave that for the next book. I'm not in the right headspace to write about it now, not while waiting for my presence in the room.

Last year, I wanted to read more authors who discuss immigration. I find affirmation to be me among pages overflowing with my imaginary friends, dead or alive.

I never thought my adventure in New York would be so challenging. New York has many stimulations and is a vast city. Sometimes, because I already know what I'll find when I climb the subway stairs because I know the paths because I have several corners of my own - sometimes, I forget that I live in the center of the world. But getting back to the point: New York has many stimulations, it's easy to lose yourself. Studies show that people who move to other countries develop more anxiety and depression. I want to avoid diving into the issue of emigration abroad; I'm speaking of internal emigration. I'm speaking of those who are immigrants of their own selves.

Once, I wrote that I wanted to increase the volume of the world so I wouldn't hear my own questions. I did that well. Being in constant change means turning the world up high. Adapting is a process that takes over everything. It's easier not to look at oneself when survival is necessary. But I'm tired. I'm still. I've been at a standstill for three years.

The rest are projects, drawers, chats at the bar, on Zoom, in bed, hangovers, mirror selfies, in the living room, in the fitting room, on the plane, and on the Uber ride back home.

The rest are the echoes, chronicles, and essays about every time I left home. If I were a novelist, this would be a road book, like in movies. It started in Rio, went to São Paulo, and came to New York, but it doesn't end here, just like movies don't simply end. Art represents a moment of everything it is to be - and to be, why not? - in this time and space. Just a fraction of a spark that was unlucky (or lucky?) to be captured and immortalized like a letter written on a deserted island and sent in a bottle, ignorant of the dangers ahead. Perhaps the letter will be found in the ocean by someone who truly needed to find it (the letter or this book).

P.S.: One day, I thought about naming this chronicle the same as the book. You can have an opinion on the matter if you want, but it's too late now.

I am from Rio de Janeiro

(This title should be read with Cássia Eller's voice - a late Brazilian singer)

This is my story and the story of many people I know. Whenever I think about these other women who connect with me through the places we have lived in, I feel like talking more about it, but I'm not sure how.

I was born in Rio de Janeiro, lived in Sao Paulo, and now live in New York. The cities I've lived in say a lot about my worldview.

Despite being very touristy, Rio de Janeiro is not cosmopolitan. The city's culture is parochial, and cariocas are proud not to expand too much to defend their territories and outdated traditions.

São Paulo is the biggest city in Brazil, where you can find everything. People from all corners of the country have multiple stories, growth ambitions, and little tradition. Are there traditional families in São Paulo? Of course. However, the city's focus is not on the past.

New York is like São Paulo multiplied by ten. It's intense, diverse, traditional, cosmopolitan, dramatic, extreme. There's no middle ground here, no shortcuts, no half-measures. Here, it's either all or nothing. It's dark, mysterious, flashy, and voracious. It's scary but enchanting. It's this frenetic contradiction of wanting everything or nothing.

And I am all of that mixed with fear and courage.

I am like this because I live these experiences, or do I live these experiences because I am like this?

That's the question.

A place is what we make of it

In 2022, returning from the end-of-year holidays in Brazil, I arrived in New York for the first time and saw the city without the filter of my expectations.

To say that New York is just like any other city is a disservice – more so to other capitals than to itself – but we must not forget that a place is what we make of it.

I wanted to move to see more of the world, experience different realities, and break free from my bubble, but exploring is scary, and the more my conscious mind called for new experiences, the more my subconscious sought comfort. So, instead of expanding, I shrunk away.

It takes time to adapt to a new city. What is called a "comfort zone" for me was the most uncomfortable moment in the world: a cold and dark place on the most vibrant summer day.

But all of that is behind me now, and I was able to see the city for all its glory and challenges without the rose-colored or gray filter. After all, a place is what we make of it, and even the brightest city in the world goes to bed early when we have no life to live it up.

Love heals, and life flows

The other day, I was talking to a close friend about the process of weaning off the anxiety medication I had been taking for almost two years. It took nine months just to discontinue the medication slowly. Entering psychiatric treatment is a trap, buddy, but sometimes it's necessary. In my case, it was necessary, and it took months to adjust the medication as well. My experience with the medication felt like a flight from Rio to São Paulo. I spent more time taking off and descending than actually in the air, or rather, in the treatment.

The conclusion of our conversation that day led to the title in question here. The phrase stuck in my head. I could always count on the unconditional love of certain people every time I needed to rise again. Love is the flickering light that weakens when nothing makes sense. Love doesn't pull you out of the depths but shows you the way out.

I often say that if I'm alive today, it's because of myself. I accepted being loved even when I thought I didn't deserve it, even without many references to what love was.

The end of existential crises

I can't remember a time in my life when I wasn't in an existential crisis. Seriously. I spent decades wanting to be someone I'm not and trying to convince myself that I could be a different person if I daydreamed enough.

Perhaps that's why it's so challenging to accept my current state of complete happiness.

Who would have thought: me? happy? My mother will be furious when she finds out I found happiness as far away from her as possible. Still, she's always angry about everything, just another instance where I lack control over the confusing emotions of my progenitor.

I used to be concerned about whether or not to write about my life. I was never encouraged to talk about myself; my life was supposed to be kept a secret among family and friends. Those who know little believe in the facade.

I began creating content seriously in December 2018 with a podcast. I sent the first episode to my parents, intending to show them that I had talents and that even though I hadn't pursued what they wanted, I was a communicator with subjective and technical skills. I received the following responses:

– You shouldn't talk so much about yourself. The episode is too long.

Both of these responses can be read as:

– Your content isn't all that interesting.

Olha que ironia: eu reclamo do ódio da minha mãe e estou aqui, destilando a minha falta de amor, sem consideração com os leitores.

Look at the irony: I complain about my mother's hatred, and here I am, spewing my lack of love without consideration for the readers.

Forgive me!

We were talking about how I overcame my existential crises. My father will say he tried to explain happiness to me all his life, even though he still doesn't quite understand it, and he will mention some psychologist he's currently obsessed with.

I, who am no longer so lost, will not try to defend myself or even talk about my accomplishments because I miss a pat on the back, a celebration, a genuine show of interest in the path I took to get here.

Other crises with different themes will come, like Christmas movies that repeat the plot but not the characters or vice versa. But I will know how to deal with the new crises with the firmness of someone who no longer wears masks.

My only constancy was writing this book

I change my mind too much. I blame it on being a Gemini ascendant, but the other day, I discovered that my twin is in house zero, which means my Gemini ascendant is almost in Taurus.

And all this Taurus map – there are other planets – makes me want guarantees (again: the fault is mine, and I place it wherever I want). Even without guarantees and with a million ideas, something in me is constant: the desire to write a book, this book, my book.

This is the only subject that has never died, one that I've talked about since the podcast, since Fotolog, and since I was eight years old. Not with so much clarity, of course! But this subject is always present, unlike the Tiffany I stopped mentioning when I got one, unlike the desire to drive that died when I had to drive for several days, unlike all the ideas that come and die at the speed of light in my mind.

The book is constant, one of the few desires that, even without guarantees, persists from my sleep to breakfast, navigating through bureaucratic tasks, walks, meditations, diaries, shows, and every conscious or unconscious second of my existence. Without it, without this, without you, I am not.

A book is perpetual, and it's scary

The other day, during a live event I was invited to, they asked me if I was afraid to release the book. I answered emphatically that I was scared of everything.

You've read this in the book, but now that it's coming to an end for both you, reading it, and myself, writing it, I think it's important to quote myself – I find it quite chic when I do this over lunch with my wife in my own book. I think it can even be sublime, even though I'm not the late Brazilian writer Fernanda Young:

– It's scary if it goes right; it's scary if it goes wrong; it's scary to be judged, to be thought crazy; for my family to say, "Why are you exposing your life so much?" It's scary to sell two copies; it's scary to sell twenty thousand copies... it's scary about everything.

The conversation continues:

– This is a non-fiction book based on memories. Memories are not journalistic facts. They are memories. They are my perceptions of what happened to me at a particular moment.

I have other fears besides the ones I mentioned on an Instagram live. I have already shared some of them throughout the book, and the last one - at least for now - is being criticized by my family, who didn't even show up at my wedding.

It didn't take me the same three seconds to write the above sentence as to find peace, but I found it and nobody will take this away from me. Not even my own shadows.

Breaking out of the cage

Once, an actress told me she couldn't write if her life depended on it. That statement tied my brain in knots.

– What do you mean someone can't write? Writing is like breathing; there is no not writing. After that day, I started noticing the activities I couldn't do if my life depended on them: singing is the most frustrating.

In a session, my psychologist asked me how I see myself in retirement, without writing. I replied that retirement is okay, but not writing is not an option.

I spent over a decade breathing through machines, listening to scripted speeches, living half-heartedly, doing everything expected of me, but not exactly. I went to rescue myself in the depths of dead-end shortcuts, abandoned arguments and learned that I could continue to escape. Escaping is a familiar pain, but I lost impunity. Escaping today is living condemned by my conscience. It's being half alive, knowing what it's like to be whole, succumbing to what is not me, and becoming a zombie to please those who cage me.

Living is a job we all need

– Life is tough!

The phrase emphatically delivered by Brazilian actor Selton Mello in the movie "Drained" (O Cheiro do Ralo) never left my mind. It became one of my mantras.

Life is tough, I sarcastically said, every time I said a "no" that should have been a "yes" in the most fragile moments of my existence.

Life is tough. My current favorite author repeats the phrase in English:

– *Life is hard...*

But the author beautifully complements the phrase:

– *...but we can do hard things.*

In difficult moments, I repeat to myself and to all my anxious, anguished, and certainty-seeking parts that will never come:

– We are capable of doing hard things.

The whimsical world of Lari – where I hid for so many years from the contacts that rubbed in my face the extent of my smallness – finally bid farewell without music and white letters on a black screen, returned to the silence of the castle illustrated in a book from my childhood, the only one someone thought to give me. Still, this memory is a topic for another book, and I've rambled on it for too long.

With permission from Glennon Doyle, we are capable of doing difficult things because living is the only job worth killing ourselves for.

Living with wholeness is the only possible job.

Milton Keynes UK
Ingram Content Group UK Ltd.
UKHW011918120724
445613UK00002B/22